LOWER DEVONIAN CONODONTS (*HESPERIUS-KINDLEI* ZONES), CENTRAL NEVADA

Lower Devonian Conodonts

(*hesperius-kindlei* Zones),
Central Nevada

by Michael A. Murphy and
Jonathan C. Matti

UNIVERSITY OF CALIFORNIA PRESS
Berkeley • Los Angeles • London

UNIVERSITY OF CALIFORNIA PUBLICATIONS IN GEOLOGICAL SCIENCES

Editorial Board: D.I. Axelrod, W.B.N. Berry, R.L. Hay, M.A. Murphy,
J.W. Schopf, W.S. Wise, M.O. Woodburne

Volume 123

Issue Date: October 1982

University of California Press
Berkeley and Los Angeles, California

University of California Press, Ltd.
London, England

ISBN: 0-520-09661-4
Library of Congress Catalog Card Number: 82-8638

Library of Congress Cataloging in Publication Data

Murphy, Michael A.
    Lower Devonian conodonts (hesperius-kindlei zones),
central Nevada.

    (University of California publications in geological
sciences; v. 123)
    Bibliography: p.
    1. Conodonts. 2. Paleontology--Devonian.
3. Paleontology--Nevada. I. Matti, Jonathan C.
II. Title. III. Series.
QE899.M87 1982   562'.2   82-8638
ISBN 0-520-09661-4   AACR2

Contents

List of Illustrations vii
List of Tables viii
Acknowledgments ix
Abstract x
Introduction 1
    Stratigraphic Framework 1
Summary 4
Systematic Paleontology
    Family Polygnathidae 6
    Family uncertain 41
    Family Icriodontidae 44
Bibliography 63
Plates 67

List of Illustrations

Figure 1 2
   Location map

Figure 2 12
   Sequence of _Ozarkodina_ _stygia_ morphs

Figure 3 14
   Phylogenetic hypothesis for the Genus _Ancryodelloides_

Figure 4 15
   Hypothesis of the ranges and relationships within
   the Genus _Ancryodelloides_

Figure 5 19
   Comparison of basal cavities in (P) elements in
   species of _Ancryodelloides_

Figure 6 32
   Hypothesis of evolution within the Genus _Amydrotaxis_

Figure 7 43
   Elements of the _Erika_ apparatus

Figure 8 46
   Conical elements of the _Pedavis_ apparatus

Figure 9 48
   Ranges and hypothesized relationships of the species
   within the Genus _Pedavis_

Figure 10 57
   Hypothesis of relationship and stratigraphic
   distribution of _Icriodus_ species in central Nevada

List of Tables

Table 1 .... following page 2
   Range Chart of Copenhagen Canyon Section V-IV
   (COP V-IV).

Table 2 .... following page 2
   Range Chart of Simpson Park Range Section VII
   (SP VII).

Table 3 .... following page 2
   Range Chart of Mill Canyon Section, Toquima
   Range (MC).

Table 4 .... pages 36-37
   Occurrence data for the elements of the
   Amydrotaxis sexidentata n. sp. and A. Johnsoni
   beta apparatuses.

Acknowledgments

Gilbert Klapper has given much valuable advice throughout the 14 years of this study. J. G. Johnson donated important material from COP II 378 collected by his students. Klapper, Johnson and W. B. N. Berry read the manuscript. Many friends and students have assisted in the field work. Among the larger contributors, we thank R. Colman, S. Finney, M. Miller, T. Morgan, and L. Schremp. Laboratory assistance was provided by W. Campbell, M. Kling, T. Morgan and J. Power. The work was supported by National Science Foundation grants EAR 76-08980, EAR 77-21624, and EAR 80-08525 and by the University of California, Riverside, Intramural Research Fund.

Abstract

This study concerns conodont faunas from the lower half of the Lower Devonian in central Nevada. It provides an enlarged taxonomic base for biostratigraphic studies of the Cordilleran region.

One new genus and nine new species furnish additional characterization of the eurekaensis, delta, pesavis, and sulcatus Zones. The genus Erika, with type species Erika divarica n. sp., and the species Ozarkodina paucidentata n. sp., Ancryodelloides omus n. sp., A. limbacarinatus n. sp., Amydrotaxis sexidentata n. sp., Pedavis biexoramus n. sp., P. brevicauda n. sp., P. breviramus n. sp., and O. pandora pi morph n. morph are described. The generic concept of Ancryodelloides is examined and expanded to include species asymmetricus, delta, eleanorae, and transitans, formerly assigned to Ozarkodina, and the evolution of the genus in the upper eurekaensis Zone from Ozarkodina through A. omus n. sp. is suggested. The Amydrotaxis apparatus reconstruction is further documented by the inclusion of data from A. sexidentata n. sp. An hypothesis for Pedavis evolution from the eurekaensis Zone through the sulcatus Zone is proposed and a modified concept of the Pedavis apparatus, one which includes six elements, is suggested. The Icriodus steinachensis lineage is expanded to include an additional morphotype and is proposed as the immediate ancestor of I. claudiae Klapper and Johnson.

Stratigraphic occurrences are provided from three sections in the region, Coal Canyon in the northern Simpson Park Range, Copenhagen Canyon in the Monitor Range, and Mill Canyon in the Toquima Range.

INTRODUCTION

The conodont faunal sequence from the lower half of the Lower Devonian was first discussed by Klapper (1969) who formulated the first complete biostratigraphic subdivision for the Lower Devonian of North America (Klapper and others, 1971, fig. 1). Many of the taxa have remained in open nomenclature until the recent publication of Lane and Ormiston (1979) on the fauna of the Salmontrout Formation of Alaska, Klapper and Murphy (1980), who discussed several of the critical zonal indices for the delta Zone, and Murphy, Matti, and Walliser (1981), who discussed the Ozarkodina remscheidensis-Eognathodus sulcatus lineage in Nevada and Germany.

The present contribution describes the remaining adequately known taxa from rocks in the interval hesperius Zone (Klapper and Murphy, 1975) through the kindlei Zone (Klapper and Johnson, 1980). The work establishes a more comprehensive base for the discussion of Lower Devonian biostratigraphy and correlation within the Cordilleran region. In addition, it documents the evolutionary history of some taxa, particularly lineages in the genera Amydrotaxis, Ancryodelloides, and Icriodus. Pedavis and Ozarkodina stygia lineages are still too poorly known to understand the details of their evolution, but these taxa have biostratigraphic utility based on their broad geographic distribution and consistent association with the better known lineages.

Stratigraphic framework. Three areas constitute the main source of material for the study (Text-fig. 1), the Coal

1

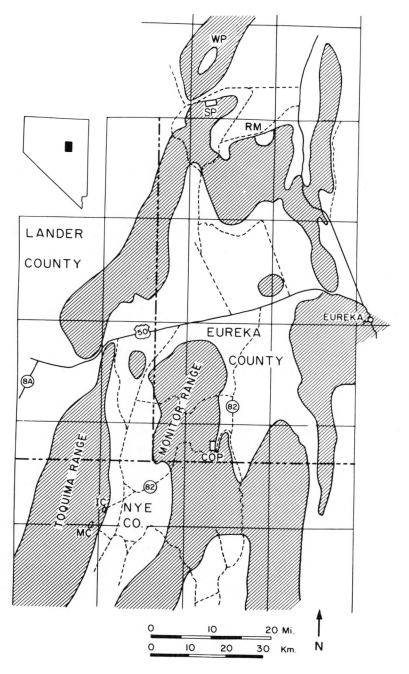

Text-figure 1. Location map.
Cross-hatched areas show mountain ranges where mea-
sured sections mentioned in this report are located.
WP - Wenban Peak, Cortez Range; SP - Coal Canyon,
Simpson Park Range; RM - Birch Creek, Roberts
Mountains; COP - Copenhagen Canyon, Monitor Range;
IC - Ikes Canyon, Toquima Range; MC - Mill Canyon -
Toquima Range.

Table 1. Range Chart of Copenhagen Canyon Section V-IV (COP V-IV).

Key:
- ∅ FRAGMENTAL OR BROKEN
- j JUVENILE
- v VICARIOUS ELEMENT
- Im INTERMEDIATE FORM

Taxa (rows):
- Icriodus woshmidti hesperius
- Ozarkodina paucidentata
- Pedavis biexoramus
- Ozarkodina eurekaensis
- Amydrotaxis sexidentata
- Ozarkodina remscheidensis
- Ozarkodina excavata
- Ozarkodina confluens group
- Ozarkodina pandora α
- Ancryodelloides omus α
- Erika divarica
- Pedavis breviramus
- Amydrotaxis johnsoni α
- Ancryodelloides omus β
- Ancryodelloides transitans
- Ancryodelloides eleanorae
- Ancryodelloides asymmetricus
- Ozarkodina stygia
- Ancryodelloides limbacarinatus
- Ozarkodina excavata var. tuma
- Ancryodelloides trigonicus
- Icriodus steinachensis H
- Ozarkodina pandora β
- Ozarkodina pandora γ
- Ozarkodina pandora δ
- Amydrotaxis johnsoni β
- Pandorinellina cf. P. optima
- Ozarkodina pandora π
- Pedavis pesavis
- Pandorinellina steinhornensis
- Pandorinellina optima
- Ozarkodina pandora ε
- Ozarkodina pandora ζ
- Icriodus steinachensis β
- Icriodus claudia
- Pedavis brevicauda
- Eognathodus sulcatus K

Column headings: UCR MUSEUM NUMBER, FEET IN SECTION, FIELD NUMBER, KILOGRAMS

Stratigraphic section markers: COP V, COP IV, COP IVA, COP IVB

Table 2. Range Chart of Simpson Park Range Section VII (SP VII).

Taxa and header rows (listed top to bottom in the chart):

- UCR MUSEUM NUMBER
- FEET IN SECTION
- Monograptus birchensis
- Monograptus microdon
- Monograptus uniformis
- Linograptus posthumus ss.
- Ozarkodina eurekaensis
- Ozarkodina remscheidensis
- Ozarkodina paucidentata
- Amydrotaxis sexidentata
- Monograptus aequabilis
- Icriodus n.sp. G. Klapper
- Monograptus praehercynicus ?
- Amydrotaxis johnsoni α
- Ozarkodina excavata tuma
- Ancryodelloides omus α
- Erika divarica
- Pedavis breviramus
- Ancryodelloides transitans
- Ancryodelloides elenorae
- Ancryodelloides delta
- Ancryodelloides asymmetricus
- Monograptus hercynicus
- Ancryodelloides limbacarinatus
- Ozarkodina stygia
- Pandorinellina cf optima
- Amydrotaxis johnsoni β
- Ancryodelloides trigonicus
- Monograptus hercynicus nevadensis
- FIELD NUMBER
- KILOGRAMS

Table 3. Range Chart of Mill Canyon Section, Toquima Range (MC).

Row categories (left margin, top to bottom):

- UCR MUSEUM NUMBER
- FEET IN SECTION
- Reworked conodonts
- *Icriodus woschmidti hesperius*
- *Ozarkodina paucidentata*
- *Amydrotaxis sexidentata*
- *Ozarkodina excavata*
- *Ozarkodina remscheidensis*
- *Icriodus* n. sp. G Klapper
- *Amydrotaxis johnsoni* α
- *Ancryodelloides transitans*
- *Ancryodelloides eleanorae*
- *Ancryodelloides delta*
- *Ozarkodina stygia*
- *Pedavis breviramus*
- *Ancryodelloides trigonicus*
- *Ancryodelloides kutscheri*
- *Amydrotaxis johnsoni* β
- *Erika divarica*
- *Ozarkodina excavata* var. *tuma*
- *Ancryodelloides asymmetricus*
- *Ancryodelloides limbacarinatus*
- *Icriodus steinachensis* H
- *Pandorinellina optima*
- *Ozarkodina pandora* α
- *Ozarkodina pandora* π
- *Ozarkodina pandora* β
- *Ozarkodina pandora* γ
- *Ozarkodina pandora* δ
- *Pandorinellina* cf. *P. optima*
- *Pedavis pesavis*
- *Ozarkodina pandora* β−ε
- *Pandorinellina steinhornensis*
- *Ozarkodina pandora* ε
- *Eognathodus sulcatus* ζ−H
- *Eognathodus sulcatus* H
- *Pandorinellina* ? cf. *P. boucoti*
- *Eognathodus sulcatus* θ−ι
- FIELD NUMBER
- KILOGRAMS

Canyon sections in the northern Simpson Park Range (Berry and Murphy, 1975), the Copenhagen Canyon sections (Matti, Murphy, and Finney, 1975), and the Mill Canyon section (Murphy, Matti, and Walliser, 1981).   Representative columnar sections may be found in these papers.

Other areas, however, have contributed important faunas that have helped our understanding of the taxonomy and stratigraphic ranges of some taxa.   These auxiliary sections are not yet as well studied.   They may provide answers to some of the unsolved relationships and tests for some of our current hypotheses.   The Nevada sections judged in need of further study are, in order of importance, Wenban Peak in the Cortez Range (WP), the Tor and McMonnigal Limestones at Ikes Canyon (IC) in the Toquima Range, and a more detailed study of the upper Roberts Mountains Formation in the Roberts Mountains (RM).   Outside Nevada, the Frankenwald localities in Germany, the U Topolů section in the Radotin Valley, Czechoslovakia and the Rauchkofel section in the Carnic Alps, Austria show some taxa in common with the Nevada sections.

Occurrence data for the MC (Mill Canyon), SP (Simpson Park Range), and COP (Copenhagen Canyon) sections are found in tables 1-3.   UCR (University of California, Riverside) is the depository for all specimens used in this study, except the type material of Bischoff and Sannemann (1958) at the Philipps Universität, Marburg.   UCR locality/specimen numbers (e. g. 8007/1) are listed for each figured specimen.

1

SUMMARY

The _Ozarkodina excavata_ and _O. remscheidensis_ lineages
are the main representatives of the genus in Nevada.
Amydrotaxis is believed to have branched from the _O.
remscheidensis_ lineage in the upper _hesperius_ Zone. Later,
in the uppermost _eurekaensis_ Zone, _Ancryodelloides_ branches
from the same lineage. Other additions concerning the genus
are: (1) The range of _O. eurekaensis_ is adjusted downward so
that it now overlaps with that of _Icriodus woschmidti
hesperius_. (2) _O. excavata tuma_ n. subsp. is a useful
indicator of the _delta_ Zone. (3) _O. pandora_ pi morph is an
additional useful morph of the _O. pandora_-Eognathodus
sulcatus lineage (Murphy, Matti, and Walliser, 1981) that is
restricted to the _pesavis_ and lower _sulcatus_ Zones. (4) _O.
paucidentata_ n. sp. is geographically widespread in the
hesperius and _eurekaensis_ Zones. (5) The _O. stygia_ sequence
of morphs in Nevada, Canada, and the Carnic Alps is not the
same. Correlation based on this sequence alone would be
unsound.

The genus _Ancryodelloides_ is expanded to include the
species _transitans_, _asymmetricus_, _delta_, and _eleanorae_,
formerly included in _Ozarkodina_, and the new species _A.
limbacarinatus_, and _A. omus_. The _Ancryodelloides_ lineage
proceeds from open to progressively more restricted basal
cavity and greater development of the lateral processes.

The _Amydrotaxis_ lineage has two early forms, _A._ n. sp. F
of Klapper and Murphy (1975) and _A. sexidentata_ n. sp.,
present in the _hesperius_ and _eurekaensis_ Zones. The

apparatus of A. sexidentata conforms to that suggested for
A. johnsoni (Klapper and Murphy, 1980).

Polygnathus pireneae is reported from Nevada in the base
of the kindlei Zone.

A new genus, Erika, with divergent denticles and bar-
like elements is described for the unusual Nevada species,
Erika divarica n. sp.

Pedavis species biexoramus n. sp., brevicauda n. sp.,
and breviramus n. sp. are named for species formerly in open
nomenclature and segments of the lineages within the genus
are suggested.

An hypothesis for a lineage from Icriodus steinachensis
to I. claudiae is suggested.

Documentation of ranges is given in tables for the
Copenhagen Canyon V-IV, Mill Canyon, and Simpson Park VII
sections.

2

SYSTEMATIC PALEONTOLOGY

The systematic section is arranged according to the classification of Klapper and Philip (1972). Most of the taxa dealt with arise from Ozarkodina and are grouped in the Family Polygnathidae. The Icriodontidae constitute most of the remainder of the stratigraphically important taxa. The Panderodontidae are not discussed here although they are common in our residues. The taxonomic position of the new genus Erika is unknown.

The nomenclature of Klapper and Philip (1971) for apparatus elements is preferred for Devonian faunas. Where not specified, text description refers to the (P) element of the apparatus.

After each species the number of specimens and number of localities at which they occurred is given, e.g. Material:26 (5).

Family Polygnathidae

One genus, Ancryodelloides, is assigned to the Polygnathidae in addition to those assigned by Klapper and Philip (1972, p. 99).

Genus Ozarkodina Branson and Mehl, 1933

For synonymy of multielement interpretation of the genus see Klapper (in Ziegler, 1973, pp. 211-212).

Two main branches of the genus, the Ozarkodina remscheidensis lineage and the O. excavata lineage provide

the main persistent and abundant taxa that are present in
Nevada.

The remscheidensis branch of the genus may have given
rise to Ancryodelloides at about the level of the
eurekaensis-delta Zone boundary and then to Eognathodus at
the beginning of the sulcatus Zone (Murphy, Matti, and
Walliser, 1980).

The details of the origin of the remaining species
placed in Ozarkodina are speculative because they appear in
the record without intermediate forms linking them to older
species. These species include O. eurekaensis Klapper and
Murphy, O. paucidentata n. sp. (=O. n. sp. E Klapper and
Murphy, 1975), and O. stygia (Flajs).

Ozarkodina eurekaensis Klapper and Murphy, 1975

1975. O. eurekaensis Klapper and Murphy, p. 33, pl. 5,
        figs. 1-17.

Remarks. This species was described by Klapper and Murphy
(1975) as the zonal name bearer for the second zone in the
Lochkovian in the Great Basin of the United States. New
data from the Birch Creek II section (Klapper and Murphy,
1975) show that O. eurekaensis overlaps the name bearer of
the next lower zone, I. woschmidti hesperius, so the
placement of the biostratigraphic zonal boundaries at BC II
in this study are different than in Klapper and Murphy. O.
eurekaensis is still unknown outside Nevada (Klapper and
Johnson, 1980, p. 408), but occurs in a distinctive
association of other less restricted forms in the
eurekaensis Zone.

Ozarkodina excavata subspecies tuma n. subsp.
Pl. 1, Figs. 3-9

Holotype. UCR specimen 8536/2.

O. excavata is a common and variable taxon that ranges
throughout the part of the section we have studied. One
form that has a separate geographic distribution and occurs
in the delta Zone, however, seems to have stratigraphic

utility. We here describe this form as a subspecies. The
name is from the Latin root, tum = swelling, alluding to the
inflation of the blade of the (P) element at the position of
the basal cavity.   In this variant, not only are the
platform lobes more expanded than usual, but also the blade
is expanded up to the base of the denticle row.   This
feature is accompanied by a large number of small even
denticles, no enlarged cusp, and little enlargement of the
anterior denticles. However, the mid-blade denticles are
widened at their bases so that their transverse dimension is
wider than their longitudinal dimension (pl. 1, figs. 3-9).
Compare the typical excavata (pl. 1, figs. 1, 2) which
occurs in the same zone.
Material: 940 (15).

Ozarkodina pandora Murphy, Matti and Walliser
pi morph new morph (P) element
Pl. 1, Figs. 10-24

Reference specimen.   UCR 8565/1, MC 16d, Toquima Range,
Nevada, 160 feet above the base of the section.

Diagnosis.   A morph of Ozarkodina pandora with non-
tuberculate platform lobes and a spear-shaped basal cavity
that occupies a little more than the posterior half of the
(P) element.

Description of the (P) element.   The blade is straight or
slightly curved. The upper profile is low except for one or
two higher denticles at the anterior end.   Denticles are
usually discrete, smaller and elliptical anteriorly, equant
and larger in the central part of the blade.   The lower
profile is nearly straight. The platform lobes and basal
cavity occupy more than half of the length of the unit, are
slightly to distinctly asymmetrical, broadly and abruptly
expanded anteriorly, and taper at a low angle to the
posterior end.   The platform lobes are unornamented.   The
basal cavity is very shallow and may extend as a tapering
furrow to the anterior end (pl. 1, fig. 10), or the walls of

the furrow may be appressed in the anterior quartile (pl. 1, fig. 19).

Remarks. The pi morph is intermediate in form between O. pandora alpha and zeta morphs Murphy, Matti and Walliser. It differs from those morphs in the shape of the platform lobes and, especially the alpha morph in having the basal cavity extend anteriorly past the mid-point of the element. It resembles the later O. linearis (Philip) in the shape and depth of the basal cavity, but has a different lateral profile and more discrete denticles.

The taxon occurs in the lower part of the pesavis Zone. Material: 55 (10).

Ozarkodina paucidentata n. sp.
Pl. 1, Figs. 25-32, 39, 40

?1964. Spathognathodus steinhornensis remscheidensis
        Ziegler. Walliser, pl. 20, fig. 26.

1975.   Ozarkodina n. sp. E Klapper and Murphy, pl. 7, figs.
        6, 9, 10.

1977.   Ozarkodina n. sp. E Klapper and Murphy. Klapper, p.
        51.

Holotype. UCR specimen 7007/1.

Type locality. Coal Canyon, northern Simpson Park Range, SP VII, 181 feet above the base of the section.

Derivation of the name. Pauci (Latin) = few + dentata (Latin) = toothed, alluding to the suppression of denticle development behind the cusp.

Diagnosis. A species of Ozarkodina characterized by no or only rudimentary development of the three or four denticles immediately posterior of the cusp, a high, conical cusp, and platform lobes that are almost circular in top view.

Description of the (P) element. This species has high anterior denticles descending to the distinctly higher cusp just posterior of the mid-section. Behind the cusp up to four denticles may be absent or very slightly developed. These are usually followed by one or two normally developed denticles somewhat smaller than the height of the anterior

denticles.   The posterior section of the blade is inflated
laterally just below the bases of the denticles so that they
arise from a narrow flattened area on top of the blade.   The
cusp is nearly circular in cross-section as is the outline
of the platform lobes.   The lower profile may be straight to
strongly   arched.   The   basal   cavity   is   asymmetrical,
constricted but open to the posterior end of the element,
and restricted to a furrow in the anterior quartile.
Remarks.   Klapper and Murphy (1975, p. 44) suggested that
the non-platform elements were so similar to those of O.
remscheidensis that the two sets of elements could not be
distinguished.   Our   material   supports   this   observation.
Samples   with   O.   paucidentata   (P)   elements   without   O.
remscheidensis   (P)   elements   contain   the   other   apparatus
elements that one would normally assign to O. remscheidensis
where it is present.

The   species   ranges   from   the   uppermost   beds   of   the
Silurian (Klapper and Murphy, 1975, table 2), entering the
section   in   the   same   sample   with   Icriodus   woschmidti
hesperius,   to   the   top   of   the   eurekaensis   Zone   (Klapper,
1977, fig. 3).
Material: 131 (11).

Ozarkodina stygia (Flajs, 1967)
Pl. 3, Figs. 1, 2, 7, 8
1967.   Spathognathodus stygius Flajs, pl. 5, figs. 12-17.
1968.   Spathognathodus seebergensis Schulze,   p. 227, pl.17,
figs. 4, 9-11.
1969.   Spathognathodus stygius Flajs. Pölsler, pl. 1,
figs. 1-6.
1973.   Ozarkodina stygia (Flajs).   Klapper, in Ziegler, ed.,
p. 249, Ozarkodina pl. 2, fig. 8.
1979.   Ozarkodina stygia (Flajs).   Lane and Ormiston, p. 57,
pl. 1, figs. 12, 13, 45, 46; pl. 2, figs. 10, 11,
20-26, 28.
1980.   Ozarkodina stygia (Flajs).   Schönlaub, pl. 2, figs.

12-14, 18, 19, 25, 26; pl. 4, 5-10, 13, 14, 18,
21-25.

Remarks. This distinctive and highly variable taxon has
been subdivided into several morphs by Lane and Ormiston
(1979, p. 58) and by Schönlaub (1980). Both papers
postulate an evolutionary change within the taxon.
Unfortunately, two different nomenclatures were adopted for
the stygia morphs in the two papers. Since non-Linnean
nomenclature has no standing or priority, we have chosen to
follow the nomenclature of Schönlaub, rather than Lane and
Ormiston, because alphabetical order of morphs corresponds
to their approximate stratigraphic appearances. As nearly
as can be determined from the stratigraphic data of Lane and
Ormiston (1979, table 1, table 2) and their plates, the
stratigraphic range of the alpha morph (their beta) of O.
stygia overlaps with the ranges of O. remscheidensis, O.
delta, and O. sexidentata n. sp. (= O. n. sp. F of Klapper
and Murphy, 1975). This would place it in the upper
eurekaensis Zone and in the lowermost delta Zone. The alpha
morph is a relatively slight modification of a rather
standard Ozarkodina (P) element which could have arisen out
of several Ozarkodina species. Characters favoring either
O. excavata or O. paucidentata n. sp. are the high cusp and
general shape and position of the basal cavity. However,
the blade of the (P) element is inflated at the base of the
denticle row in both these species, a character not present
in O. stygia. O. remscheidensis, O. eosteinhornensis, and
O. repetitor are the other species showing similar morphologies and appropriate stratigraphic position. The O. stygia
alpha morph is clearly the most primitive and clearly the
lowest stratigraphically (Lane and Ormiston 1979, p. 58,
Schönlaub, 1980, p. 39, fig. 21). In Austria an orderly
progression from O. stygia alpha through O. stygia delta is
reported by Schönlaub. However, in the western North
American sections in Alaska and Nevada this progression is
not obvious and there seems to be an almost complete overlap
in the ranges of the post-alpha morphs (Text-fig. 2).

However, it would appear that <u>O. stygia</u> morphs will be biostratigraphically useful for subdividing the <u>delta</u> Zone as they become better known.

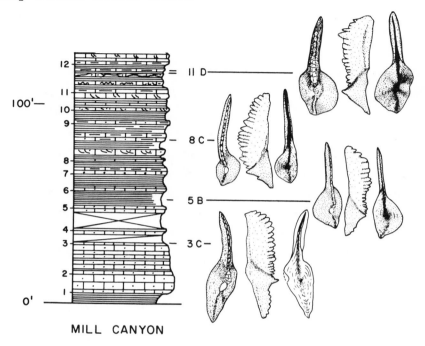

MILL CANYON

Text-figure 2.   Sequence of <u>Ozarkodina stygia</u> morphs in the Mill Canyon section, Toquima Range, Nevada. Numbers are field numbers painted on sample localities.

Genus _Ancryodelloides_ Bischoff and Sannemann, 1958

According to Bischoff and Sannemann (1958, p. 91) the characteristic features of the genus are the shape and number of the lobes on the (P) element, the smooth surface of the platform, and the narrow basal groove. The members of the genus commonly have a granular surface texture on the platform contrary to the statement of Bischoff and Sanneman. This is visible in their plates and has been noted on all other species of the genus except _A. omus_. They included two species in their concept of the genus, _A. trigonicus_, the type species, and _A. kutscheri_. For the reasons listed below, we include additionally _A. transitans_ (Bischoff and Sannemann), _A. asymmetricus_ (Bischoff and Sannemann), _A. delta_ (Klapper and Murphy), and _A. eleanorae_ (Lane and Ormiston). Subsequent to Bischoff and Sannemann, a few papers report these species in small numbers. Among these papers, Schulze (1968, p. 183), Carls (1969, p. 342) and Lane and Ormiston (1979, p. 49) are important because they reiterate the observation of Bischoff and Sannemann (1958, p. 92), who noted the close relationship between the morphs of _Ancryodelloides_ _trigonicus_, _A. transitans_ (Bischoff and Sannemann) and _A. asymmetricus_ (Bischoff and Sannemann). Lane and Ormiston also point out the close relationship of _A. transitans_ to _A. eleanorae_ (Lane and Ormiston), and _A. delta_ (Klapper and Murphy).

Additionally we place _A. omus_ n. sp. and _A. limbacarinatus_ n. sp. in the genus.

Four reasons appear to require this classification: 1) phylogenetic interpretation of the taxa, 2) the similarity of the basal cavity and groove, 3) the appearance of a distinctive type of (O) element which is apparently shared by all the _Ancryodelloides_ (P) elements, and 4) specimens with intermediate morphologies between several of the taxa.

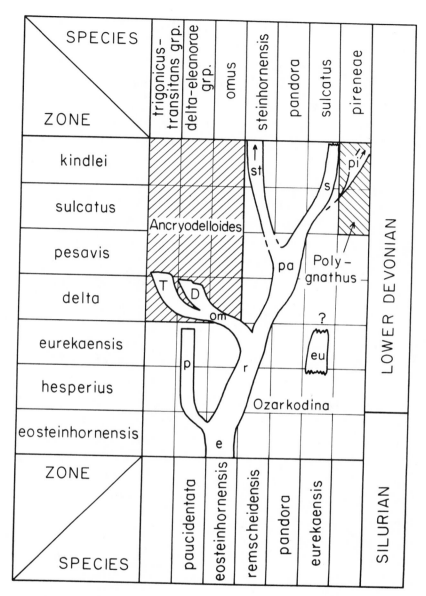

Text-figure 3.   Phylogenetic hypothesis for Genus
Ancryodelloides which regards Ozarkodina remscheidensis as the ancestral stock. Patterns indicate the
morphological space of the three genera depicted in
the diagram.   The vertical lines separate the
morphological space of the species-rank taxa.

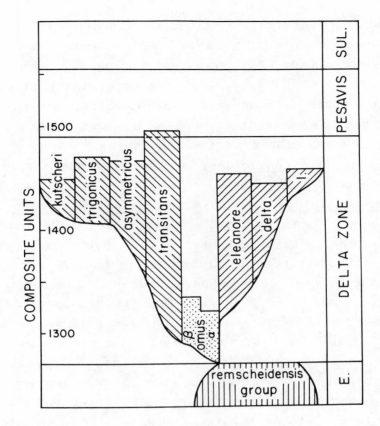

Text-figure 4.   Hypothesis of the ranges and
relationships within the Genus Ancryodelloides.   Age
is given in terms of Zones on the right.
l. = limbacarinatus; E. = eosteinhornensis Zone;
Sul. = sulcatus Zone.

It seems more logical to separate all the taxa mentioned above (except A. omus) from Ozarkodina because they differ from (P) elements of Ozarkodina in the character of the basal cavity and groove, the character of the platform, and probably in some elements of the apparatus, but the latter has not yet been worked out. Also, the species listed above are closely associated in time and space. A. omus is also assigned to Ancryodelloides, because it is interpreted as belonging to the lineage.

Ancryodelloides evolves from Ozarkodina remscheidensis (Text-figs. 3 and 4). This distinctive genus apparently begins in the upper eurekaensis Zone with the development of shouldered and flattened platform lobes (pl. 2, figs. 18-20) followed slightly later by development of tubercles on the lobes (pl. 2, figs. 21-29). This is accompanied by the development of a slight indentation that follows the base of the (P) element as described above. A further constriction of the basal cavity of either the tuberculate or nontuberculate forms gives the typical development of the pit in the genus Ancryodelloides (Text-fig. 5). The intermediate stages between O. remscheidensis and Ancryodelloides are found in A. omus n.sp. and for this reason A. omus is assigned to Ancryodelloides. Hypothesis of descent is shown in Text-figs. 3, 4.

Revised diagnosis. A polygnathid genus with a well-developed, shelf-like platform, with or without lateral processes and with a basal cavity that is more restricted than the platform except in the most primitive forms.

Ancryodelloides omus n. sp.
Pl. 2, Figs. 14, 18-29

1958.   Spathognathodus fundamentatus Walliser. Bischoff and
        Sannemann, pl. 14, figs. 1, 2.

1971.   Spathognathodus aff. transitans Bischoff and
        Sannemann. Bultynck, p. 31, pl. 1, figs. 1, 2, 4;
        ? figs. 8-13.

Holotype.   UCR 6249/6, Copenhagen Canyon Section IV, 130
feet above the base.
Derivation of the name. Om (G) = shoulder, alluding to the
shouldered platform lobes.
Diagnosis.   A species of Ancryodelloides with shouldered
platform lobes and an open basal cavity.
Description.   The (P) element has almost parallel upper and
lower profiles. The blade is sturdy with more or less
equal-sized, elliptical denticles. The posterior two or
three descend to the lower profile. A slightly indented
band follows the base of the element from just behind the
anterior edge to the posterior tip including the margins of
the platform lobes. The posterior quarter of the blade is
bent laterally. The platform lobes are shouldered and may
be surmounted by tubercles (beta morph) or may be smooth
(alpha morph). The basal cavity is open under the platform
lobes and extends posteriorly to the tip of the unit as a
deep, wide furrow. Anteriorly it is maintained as a narrow,
open furrow almost to the anterior end.
Remarks.   The Nevada examples are very similar to Spathognathodus fundamentatus Walliser (Bischoff and Sannemann,
1958, pl. 14, figs. 1, 2) from Germany, particularly in
lateral view and in the subquadrate outline of the platform
lobes. The main difference from the German specimens and
also from Spanish specimens (Bultynck, 1971, pl. 1, figs. 1,
2, 4), is the consistently more sharply bent posterior
process of the Nevada specimens. A. omus n. sp. differs
from A. transitans in having an unconstricted basal cavity.
    The (P) element stages between O. remscheidensis, A.
omus alpha and A. omus beta are found in sequence in the COP
IV section, but only a few specimens have been recovered.
However, A. omus is also abundant in bed 15 of the U Topolů
section in the Radotín Valley, Czechoslovakia (Chlupáč et
al. 1972, p. 127). The material from bed 15 shows a wider
range of variability than the Nevada collections and
includes specimens with both tuberculate and non-tuberculate
platforms.   The character of the apparatus is in doubt,

Text-figure 5. Comparison of basal cavities in (P) elements in species of Ancryodelloides.

a. A. omus n. sp., specimen figured in pl. 2, figs. 28, 29.

b. A. omus n. sp., specimen figured by Bischoff and Sannemann (1958, pl. 14, fig. 3), Philipps Universität, Marburg.

c. A. transitans (Bischoff and Sannemann), MC 5A, Mill Canyon, Nevada.

d. A. transitans (Bischoff and Sannemann), specimen figured in Bischoff and Sannemann (1958, pl. 13, fig. 4), Philipps Universität, Marburg.

e. A. transitans (Bischoff and Sannemann), MC 5B, Mill Canyon Nevada.

f. A. asymmetricus (Bischoff and Sannemann), MC 5B, Mill Canyon, Nevada.

g. A. trigonicus Bischoff and Sannemann, specimen figured in Bischoff and Sannemann, (1958, pl. 12, fig. 9). Note the difference in shape of the lateral terminations of the basal cavity between small A. trigonicus and small A. transitans (figs. e, h), Philipps Universität, Marburg.

h. A. transitans (Bischoff and Sannemann), MC 5A, Mill Canyon, Nevada.

i. A. eleanorae (Lane and Ormiston), specimen figured on pl. 4, figs. 4-6, MC 3C, Mill Canyon, Nevada.

j. A. delta (Klapper and Murphy), specimen figured on pl. 4, figs. 16-18, MC 3, Mill Canyon, Nevada.

k, l. A. limbacarinatus n. sp. MC 6, Mill Canyon, Nevada.

m. A. trigonicus Bischoff and Sannemann, holotype, figured in Bischoff and Sannemann (1958, pl. 12, fig. 12). Philipps Universität, Marburg.

n. A. kutscheri Bischoff and Sannemann, specimen figured in Bischoff and Sannemann (1958, pl. 12, fig. 17), Philipps Universität, Marburg.

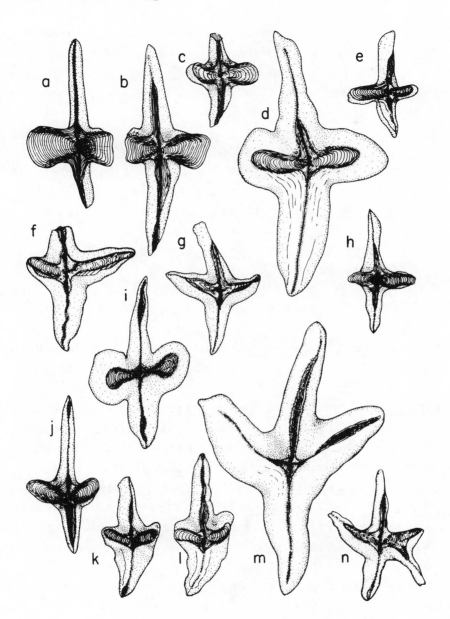

although the distinctive (O) elements with expanded basal
cavity and shouldered blade believed by Lane and Ormiston
(1979) to be (O) elements of A. eleanorae and very similar
to "Ozarkodina gaertneri Walliser," are first encountered in
beds within the range of A. omus.
Material: 24 (4).

Ancryodelloides trigonicus Bischoff and Sannemann, 1958
Pl. 3, Figs. 3-6, 11

1958. Ancryodelloides trigonica Bischoff and Sannemann, p.
       92, pl. 12, figs. 9, 12-14, 16.
1968. Ancryodelloides trigonica Bischoff and Sannemann.
       Schulze, p. 183, pl. 16, figs. 4a, 4b.
1979.   Ancryodelloides trigonica Bischoff and Sannemann.
       Lane and Ormiston, pp. 48, 49, 52, pl. 2, figs. 16,
       17.
1980. Ancryodelloides trigonicus Bischoff and Sannemann.
       Schönlaub, pl. 4, fig. 15, pl. 5, fig. 14; not pl.
       7, fig. 6.

Diagnosis. A species of Ancryodelloides in which the basal
cavity is restricted to a small pit beneath the cusp in
adult specimens. In juvenile specimens the distal ends of
the basal cavity on the lateral processes are pointed.

Remarks. A. trigonicus differs from A. kutscheri only by
the presence in A. kutscheri of a bifurcation of one lateral
process. They are probably morphs of a single species, but
are so rare that adequate variation studies have not been
possible.

Several authors (Bischoff and Sannemann, 1958; Schulze,
1968; Lane and Ormiston, 1979) have noted the close
relationship of A. trigonicus with A. transitans and A.
asymmetricus which may be additional morphotypes of this
species. However, some differences in the shape of the
basal cavity, the expansion of the platform of the posterior
process, and the angle at which the lateral processes meet
the main axis of the unit may be catalogued. In typical,
mature A. trigonicus, the basal opening is constricted to a

pit with its walls tightly appressed beneath the posterior
and lateral processes away from the pit. An open, narrow
groove is present on the anterior process (pl. 3, fig. 11).
This is accompanied by an expansion of the platform on the
posterior process, and the lateral process meets the axis of
the unit at an angle less than 90 degrees.

In typical mature A. transitans, the basal opening is
restricted (but still open) on the lateral processes
(Text-fig. 5), the platform on the posterior process is
poorly developed, and the lateral processes meet the axis of
the unit at 90 degrees.

There is variation with respect to all three of these
features, especially with respect to growth stages.

The basal opening in A. trigonicus is more open in
smaller specimens than in larger ones and may be as expanded
under the lateral processes as it is in A. transitans,
differing only in shape. The pit in A. trigonicus tapers to
a point laterally, whereas in A. transitans the basal cavity
is club-shaped and bluntly rounded laterally (Text-fig. 5).
This difference seems to be the most reliable criterion for
separating the two taxa.

The platform of the posterior process of A. trigonicus
is broad and flat and meets the blade at nearly 90 degrees.
Although the platform width shows considerable variation,
its bench-like shape is constant. In some specimens of A.
transitans, the platform of the posterior process has the
bench-like shape, but it is generally narrower than in
trigonicus. Other specimens of A. transitans have an
inflated posterior process below the denticle row, but no
shoulder or bench is developed. Young specimens uniformly
have the latter type cross section so this character varies
between mature individuals as well as with growth stage.

The angle of the lateral processes and the main axis of
the (P) element shows many intermediate stages between
trigonicus and transitans and is probably the least reliable
criterion for separating the taxa.

Material: 71 (8).

Ancryodelloides kutscheri Bischoff and Sannemann, 1958
Pl. 3, Fig. 20

1958. Ancryodelloides kutscheri Bischoff and Sannemann. P. 93, pl. 12, figs. 15, 17, 18.

1968. Ancryodelloides kutscheri Bischoff and Sannemann. Schulze, p. 183, Text-fig. 11.

1969. Ancryodelloides kutscheri Bischoff and Sannemann. Pölsler, p. 404, pl. 1, figs. 22, 23.

1979. Ancryodelloides kutscheri Bischoff and Sannemann. Lane and Ormiston, pp. 48, 49, 52. pl. 2, figs. 14, 15, 18, 19.

Remarks. See discussion under A. trigonicus.
Material: 7 (4).

Ancryodelloides transitans (Bischoff and Sannemann, 1958)
Pl. 2, Figs. 9-11; Pl. 3, Figs. 9, 10;
see also Pl. 4, Figs. 10-12

1958. Spathognathodus transitans Bischoff and Sannemann, p. 107, pl. 13, figs. 4, 5, 12, 14.

1962. Spathognathodus transitans Bischoff and Sannemann. Walliser, p. 284, fig. 1, no. 36.

1968. Spathognathodus transitans Bischoff and Sannemann. Schulze, pl. 16, figs. 1-3.

1980. Ozarkodina transitans (Bischoff and Sannemann). Schönlaub, pl. 2, figs. 22, 24; pl. 4, figs. 11, 16, ?17; pl. 5, figs. 8, 10-13; pl. 7, fig. 4.

1980. Ozarkodina limbacarinata Klapper and Murphy (nomen nudum) Schönlaub, pl. 5, figs. 1

?1980. Ozarkodina cf. eleanorae Lane and Ormiston. Schönlaub, pl. 20, fig. 4.

Remarks. See discussion under A. trigonicus.
Material: 143 (37).

Ancryodelloides asymmetricus
(Bischoff and Sannemann, 1958)
Pl. 2, Figs. 7, 8

1958.   Spathognathodus asymmetricus Bischoff and Sannemann,
         pl. 12, fig. 19; pl. 13, fig. 1.
1968.   Spathognathodus asymmetricus Bischoff and Sannemann.
         Schulze, pl. 16, fig. 10.
?1969.   Spathognathodus cf. asymmetricus Bischoff and
         Sannemann. Carls, pl. 2, fig. 20.
1973.   Ozarkodina asymmetrica (Bischoff and Sannemann).
         Klapper, in Ziegler, I:pl. Ozarkodina 1, fig. 1.

Diagnosis.   A species of Ancryodelloides with one short
non-denticulate process and one long denticulate process.

Remarks.   Schulze's figured specimen (1968, pl. 16, fig. 10)
is a juvenile, but has the long and the short process with
and without denticles, respectively.   The biconvex configuration of the normal adult upper view shown in Bischoff and
Sannemann (1958, pl. 12, fig. 19; pl. 13, fig. 1) is a
variable character in sample SP VII 359 feet.   Schulze's
specimen is easily within the range of this variability.

Material: 27 (9).

Ancryodelloides eleanorae (Lane and Ormiston, 1979)
Pl. 4, Figs. 4-6; see also Figs. 10-12

1979.   Ozarkodina eleanorae Lane and Ormiston, p. 55, pl. 1,
         fig. 40, not fig. 47; pl. 2, figs. 6, 7; pl. 3, figs.
         7, 8, 11, 12.
1980.   Ozarkodina eleanorae Lane and Ormiston. Klapper and
         Murphy, fig. 4, nos. 1, 7-9, 13, 14.
1980.   Ozarkodina delta Klapper and Murphy. Schönlaub,
         pl. 4, figs. ?20, 27-29.
1980.   Ozarkodina cf. asymmetrica (Bischoff and Sannemann).
         Schönlaub, pl. 4, figs. 3, 4.

Diagnosis.   A species of Ancryodelloides in which the (P)
element has rounded subequal platform lobes.

Remarks.   Klapper and Murphy (1980, p. 500) have discussed
the characterizing features of A. delta and A. eleanorae.

The (O) element assigned to A. eleanorae by Lane and Ormiston (1979, pl. 1, fig. 47) is commonly found with species of the genus at Nevada localities, but there is a second (O) element that we believe is more closely similar in morphology to the A. eleanorae (P) element (pl. 4, figs. 1-3). It has the closely appressed denticles and the platform extends anteriorly and posteriorly as a narrow shouldered shelf. As suggested in the generic discussion, we believe the (O) elements of A. eleanorae and A. delta to be shared. The (O) element figured by Lane and Ormiston is here assigned to A. transitans and the remaining Ancryodelloides species (e.g. pl. 3, figs. 18, 19, 21-24). All these species have a more rounded platform margin and fewer, more discrete denticles which are characters similar to the specimen of Lane and Ormiston.

According to the data in Lane and Ormiston (1979, p. 51), this species is confined to the delta Zone, as it is in Nevada, if the synonymy listed in Klapper and Murphy (1980, p. 499) is accepted. Although the ranges of A. eleanorae and A. delta are almost identical, we have not seen any intermediate morphotypes among the (P) elements.
Material: 76 (20).

Ancryodelloides delta (Klapper and Murphy, 1980)
Pl. 4, Figs. 7-9, 14-18
For prior synonymy see Klapper and Murphy, 1980, p. 499.
1980. Ozarkodina limbacarinata Klapper and Murphy.
       Schönlaub, pl. 5, fig. 4 (nomen nudum).
Remarks. This taxon has been discussed in detail recently by Klapper and Murphy (1980, p. 499, 501-502). We have changed the generic assignment of Klapper and Murphy because of the stronger morphologic relationship to Ancryodelloides taxa mentioned above under the generic description.

Although a wide range of variability in platform shape exists in A. delta, we have had no difficulty in separating it from its congeneric taxa, especially A. eleanorae. The more asymmetrical lobes and the position of the larger lobe

on the concave side of the blade in A. delta rather than the
convex side as it is in A. eleanorae may be used for
distinguishing the two taxa, in addition to the criteria
listed by Klapper and Murphy (1980, p. 501).
Material: 84 (16).

Ancryodelloides limbacarinatus n. sp.

Pl. 4, Figs. 20-30

?1980. Ozarkodina limbacarinata Klapper and Murphy.
        Schönlaub, pl. 5, fig. 3, (nomen nudum).

Derivation of the name. Limba + carinatus, Latin = marginal
carina. Refers to the parapet-like turned-up outer edge of
one platform lobe.

Holotype. UCR 8535/1, specimen illustrated on pl. 4, figs.
20, 25, 26, MC 5B Bastille Limestone, Toquima Range,
Nevada.

Diagnosis. (P) elements of A. limbacarinatus n. sp. are
characterized by a parapet that follows the outermost margin
of the outer platform lobe in mature individuals (0.8 mm or
more). It also has a basal cavity that is restricted to the
area under the lobes of the platform.

Description. The platform occupies a little more than half
the element. The shouldered platform lobes are broadly
expanded at mid-length and taper irregularly to a point
posteriorly. The two lobes extend about equally anteriorly.
The outer lobe is wider and bent upward to form a parapet
along its outer margin. The inner lobe generally is flat,
warped, or undulose, but some specimens show a narrow,
raised area around the platform margins and the depression
of the upper surface as is found in its predecessor, A.
delta. Both lobes have a granulose surface texture.

    In lateral aspect, the profile is typical of some other
members of the genus with high anterior denticles, low
mid-blade denticles, and moderate posterior denticles. The
lower profile is variable, but generally concave. The
denticles are elliptical and free at their tips. The basal
cavity is shallow below the lobes and is less extensive than

the lobe.   The lateral terminations of the cavity are
rounded (Text-fig. 5). Generally, anterior and posterior of
the basal cavity, the walls of the blade are tightly
appressed and the furrow is inverted.

In lower view the outer platform commonly has a low,
rounded ridge trending obliquely to the rear.

Remarks. A. limbacarinatus n. sp. is distinguished from all
other species of Ancryodelloides by the parapet on the outer
margin of the outer lobe.   It is surely derived from A.
delta which it resembles most closely in platform shape,
surface configuration, and surface ornamentation.   Large
specimens of A. delta may show the inverted basal furrow of
A. limbacarinatus.

The species is narrowly confined stratigraphically and
is commonly accompanied by Ozarkodina excavata tuma n.
subsp. Its range overlaps with A. delta.

Material: 136 (11), including material from Wenban Peak
section.

Genus Pandorinellina Müller and Müller, 1957

Type species. Pandorinellina insita (Stauffer, 1940)

According to Klapper (1973, p. 317), the possession of a
diplododellan (A$_3$) element distinguishes Pandorinellina
species from those of Ozarkodina.   The presence of the
(A$_3$) element with a posterior process (diplododellan) is
extremely rare in our collections and we have found
relatively few platform elements assignable to the genus.
The species assigned to Pandorinellina previously in the
Nevada Lochkovian-Pragian are P. optima? and P.
steinhornensis miae (Klapper and Johnson, 1980, tables 2,
5).   Pandorinellina? boucoti (Klapper) is also present in
our collections.   It has been suggested by Murphy, Matti,
and Walliser (1981, Text-fig. 4) that the steinhornensis
group originated near the top of the pesavis Zone from
Ozarkodina pandora Murphy, Matti and Walliser in contrast to

the classification of Klapper (1977) who places steinhornensis in Pandorinellina (Text-fig. 3).

Pandorinellina cf. P. optima (Moskalenko, 1966)
Pl. 2, Figs. 1-3

Diagnosis. A species of Pandorinellina with three or four high anterior denticles followed by a small and crowded denticle and then a series of about five or six more or less uniform denticles of the same height with several shorter denticles at the posterior end. It has a straight posterior lower profile.

Description. In addition to the characteristics in the diagnosis, the basal cavity is approximately at mid-length with anterior and posterior basal furrows extending to the distal ends of the processes. The denticles are flattened or elliptical. The anterior lower profile ranges from convex to concave and is paralleled by a constricted band on the lower one-third of the blade. There is a tendency for the anterior denticles to be wider but not higher than the mid-blade denticles. The asymmetrical platform lobes are rounded subquadrate or trigonal.

Remarks. P. cf. P. optima differs from the typical P. optima from the Zervashan Range in being shorter, having fewer denticles posterior to the high anterior group (8 as opposed to 12), and the posterior lower profile is straight rather than concave. Its stratigraphic position is approximately the same as the Zervashan material (Mashkova, 1972, pl. 2, figs. 7-12; Klapper and Philip, 1972, pl. 1, figs. 1-11) which is listed as coming from the Monograptus hercynicus Zone.

Pandorinellina optima (Moskalenko, 1966)
Pl. 4, Figs. 13, 19

1966. Spathognathodus optimus Moskalenko, pp. 88, 89, pl. 11, figs. 12-15.

1969. Spathognathodus optimus Moskalenko. Klapper, p. 20, pl. 4, figs. 13-29.

1972.   Ozarkodina steinhornensis optima (Moskalenko).
        Mashkova, p. 84, pl. 2, figs. 7-12.
1972.   Pandorinellina optima (Moskalenko). Klapper and
        Philip, p. 99, pl. 1, figs. 1-11.
1973.   Pandorinellina optima (Moskalenko). Klapper in
        Ziegler ed., Ozarkodina pl. 2, fig. 12.

Remarks.   The Nevada specimen (pl. 4, fig. 13) is most
similar to the Canadian form figured by Klapper (1969, pl.
4, figs. 22, 23) from the pesavis Zone.   In that specimen,
there is a marked diminution in size of the denticles in the
center of the blade, but the Canadian specimen does not have
a strongly angular lower profile.   A second Nevada specimen
(pl. 4, fig. 19) is close to a second Canadian form
(Klapper, 1969, pl. 4, fig. 19).   In a number of other
Nevada specimens (pl. 2, figs. 4-6), the biconvex profile
and concave posterior lower profile are developed, but also
one or two strongly developed denticles are found above the
basal cavity and the specimens have the platform shape of O.
remscheidensis (Ziegler).   We have identified specimens with
these characteristics as O. remscheidensis.

Pandorinellina? cf. P.? boucoti (Klapper, 1969)
Pl. 3, Figs. 12-17

1969.   Spathognathodus boucoti Klapper, p. 15, pl. 6,
        figs. 1-8.
1980.   Pandorinellina? boucoti (Klapper). Klapper and
        Johnson, p. 540.

Remarks.   Pandorinellina? boucoti (Klapper) is a rare
species previously reported only from Royal Creek in Yukon
Territory, Canada.   Klapper illustrates two morphologies in
his paper, that of the holotype from the pesavis Zone and
that of a larger specimen from the sulcatus Zone.   The
latter is accompanied by specimens similar to the holotype
and so it is inferred that the differences between the two
morphologies result from differences in growth stage.
Specimens similar to the holotype occur in the pesavis Zone
at Coal Canyon in Nevada.   Specimens more like the larger

specimen figured by Klapper (1969) are present in the Tor
Limestone and in the Bastille Limestone in the Toquima
Range, Nevada, but these specimens all have a pillar-like
growth rising from both platform lobes up to the level of
the denticles. The growth above the wider platform lobe
ends in an irregular double node. The one above the narrow
platform lobe is double or single noded. Above the position
of the basal cavity the blade is broadened and has a pit-
like sulcus between the posterior row of denticles and a
large, high, triangular, anterior denticle. These specimens
are classified here as Pandorinellina? cf. P.? boucoti
(Klapper).

Genus Amydrotaxis Klapper and Murphy, 1980
Type species. Spathognathodus johnsoni Klapper, 1969.

Klapper and Murphy (1980) erected this genus to house a
small group of species from the Cordilleran region of North
America. The genus is characterized by a type 2 apparatus
(classification of Klapper & Philip, 1971, p. 433), but with
strongly compressed denticles, a basal cavity that is well
developed in all elements, and a symmetry transition series
in which the basal cavity is expanded beneath the cusp and
constricted away from the cusp. The (P) element has thick
walls with a small number of low denticles.

The blade of the (P) element is slightly curved both to
the right and to the left in upper view. For this reason we
do not employ the terms "inner" and "outer" sides of the
blade. Instead, following Klapper and Murphy (1980), we
substitute the terms wider and narrower side of the
platform.

Amydrotaxis has not been reported from Europe and was
not known from arctic North America until recently (Lane and
Ormiston, 1979). Three species are currently tentatively
included in the genus. Amydrotaxis johnsoni (Klapper), the
type species, and A. sexidentata n. sp. are restricted to

the cordilleran region of North America. A. druceana has been reported from Australia (Pickett, 1980).

The genus arises in the eurekaensis zone (middle Lochkovian), probably ultimately from O. remscheidensis.

The Amydrotaxis Apparatus

The Amydrotaxis apparatus has been discussed by Pickett (1980) for A. druceana and by Klapper and Murphy (1980) for A. johnsoni morphs. Here we present, in addition, the locality and data for the morphs of A. johnsoni beta and for A. sexidentata in table 4 and further documentation of the apparatus reconstruction.

For the 36 samples from which A. sexidentata was identified, 16 contain the platform element and at least part of the apparatus. Three collections from the Lone Mountain Dolomite contain only elements of the apparatus. At two of the remaining localities the total number of conodonts recovered was very low, so it is not surprising that these two localities do not show the association required by our reconstruction.

At the remaining three localities the maximum number of inferred Amydrotaxis elements present in the sample is two, even when other conodont elements were relatively abundant (in one case 95 specimens). This indicates to us that the A. sexidentata individuals were numerically a minor part of the assemblage and that, again, it is not surprising that the predicted association is not present in the samples. We, therefore, regard these statistics as supporting the reconstruction that was suggested for A. sexidentata by Klapper and Murphy (1980, p. 497).

Amydrotaxis sexidentata n. sp.
Pl. 5, Figs. 1-27

1975. Ozarkodina n. sp. F Klapper and Murphy, in part, not pl. 5, figs. 24-26.

1977. Ozarkodina n. sp. F Klapper and Murphy. Klapper, p. 38.

1979.   Ozarkodina n. sp. F Klapper and Murphy. Lane and
        Ormiston, pl. 1, figs. 6, 7, 9, 11.

Derivation of the name. Sexi (Latin) = six + dentat (Latin)
= toothed: refers to the usual number of denticles born on
the blade.

Holotype.   UCR 8524/1, bed at MC 2G + 7cm, Toquima Range,
Nevada.

Diagnosis.   A. sexidentata is characterized by the
co-occurrence of a low number of denticles on a blade that
tapers to a point at the posterior end and a broadly
expanded, slightly posterior, asymmetrical platform without
sulcate indentation of the platform lobes in lower view.

Description. (P) element. The blade is almost straight to
slightly curved with the posterior tip of some specimens
bent at a low angle. The blade usually has a low, even or
posteriorly descending upper profile and almost straight
lower profile. The upper profile tapers at the posterior
end at 30 to 60 degrees to meet the lower profile at a
point.   The denticles are broad, discrete, poorly
differentiated below the crest of the blade, and variable
with respect to the position of the larger or higher ones.
They commonly number six, but five to seven have been
observed. The posterior-most denticle is often enlarged and
inclined to the rear.

The platform is moderately to strongly asymmetrical with
the larger lobe either on the slightly convex or the
slightly concave side of the blade. The lobes may extend
outward at right angles to the blade or be inclined slightly
anteriorly.   The apex of the shallow basal cavity is
slightly posterior of the center and extends posteriorly as
a very shallow trough to the end of the element. Anteriorly
it is an open, narrow groove.

Remarks.   A. sexidentata n. sp. differs from all the
morphotypes of A. johnsoni in the shape of the narrower
platform lobe.

A. sexidentata is interpreted to be the direct ancestor
of A. johnsoni alpha morphotype (Text-fig. 6) which it

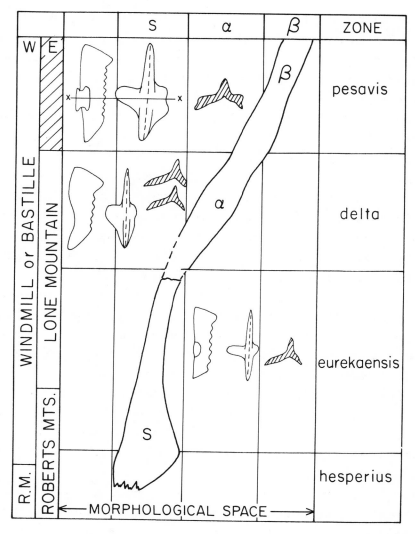

Text-figure 6. Hypothesis of evolution and ranges within the genus <u>Amydrotaxis</u> in North America. Cross-hatched figures show cross sections through the basal cavity of the (P) element along line x-x. S = <u>sexidentata</u>, alpha and beta = morphs of <u>johnsoni</u>. More westerly (W) and more easterly (E) formations are depicted in the left columns.

precedes stratigraphically both in Nevada and Alaska.   Lane
and Ormiston (1979, pl. 1, figs. 9, 11) have figured a
specimen that is transitional between the species in having
an enlarged basal cavity that tapers gradually to a point
posteriorly as in A. johnsoni alpha.   The narrow lobe,
however, is not sulcate as in johnsoni and the denticles
number only six, the normal number in sexidentata (see also
pl. 5, figs. 3, 4, herein).   A. sexidentata also seems to be
closely related to O. n. sp. F Klapper and Murphy.   O. n.
sp. F differs in the shape of the platform and the development of two or more very large, triangular denticles.

Several localities in the Lone Mountain Dolomite,
northern Roberts Mountains have recently yielded small
numbers of specimens of the A. sexidentata n. sp.
apparatus.   The (P) element is the only (P) element in the
sample and the other elements agree with the reconstruction
of Klapper and Murphy (1980) for the genus. These samples
and the ones for A. johnsoni reported earlier (Klapper and
Murphy, 1980, p. 503; Matti et al., 1975, p. 36) from the
Lone Mountain Dolomite of the Sulphur Springs Range indicate
that members of this genus ranged into shallow water
habitats where other polygnathid conodonts were rare or
absent and that they were relatively common in these
locales.   They are also common in the turbidite deposits
derived from the Toiyabe Ridge (Matti and McKee, 1977) at
Mill Canyon, Toquima Range and in Coal Canyon sections,
northern Simpson Park Range.
Material: 97 specimens of the (P) element from 31
localities; 104 non-platform elements from 21 localities
(table 4a).

Amydrotaxis johnsoni (Klapper, 1969)
1969.   Spathognathodus johnsoni Klapper, pp. 18-19, pl. 5,
        figs. 8-16.
1977.   Ozarkodina johnsoni (Klapper).   Klapper, pp. 40, 52.
1977.   Ozarkodina n. sp. C Klapper, pp. 40, 52.

not 1979. Ozarkodina johnsoni (Klapper). Lane and
        Ormiston, p. 56, pl. 3, figs. 4, 6 [=
        Ancryodelloides delta (Klapper and Murphy)].
1980. Amydrotaxis johnsoni (Klapper). Klapper and Murphy,
        1980, p. 497, fig. 2, nos. 1-19; fig. 3, nos. 1-20.

    Amydrotaxis johnsoni (Klapper) was revised by Klapper
and Murphy (1980) to include two morphotypes designated
alpha and beta (for the earlier and later forms, respectively). Each of these morphotypes has stratigraphical
significance, but intergrades morphologically with the other
morph so that samples in the transition horizons show a
range of variation including both morphotypes. It was for
this reason that Klapper and Murphy chose to represent the
group of morphotypes as a single evolving species, thus
emphasizing their relationship, rather than to adopt a
classification and nomenclature that would obscure this
relationship. Their conclusions are adopted here.

                    alpha morphotype
1971. Spathognathodus n. sp. C Klapper, in Klapper and
        others, p. 288.
1977. Ozarkodina n. sp. C Klapper, Klapper, p. 52.
1980. Ozarkodina johnsoni alpha morphotype Klapper and
        Murphy, fig. 2., nos. 1-3, 5-14, 16; fig. 3, nos.
        1-20.

Diagnosis. The (P) element. A. johnsoni alpha morphotype
Klapper and Murphy is characterized by the shape of its
platform lobes. The transversely narrower lobe has a
sloping or angulate transverse profile as opposed to a
shouldered one (beta morphotype) (Text-fig. 6) and tapers
posteriorly with or without a lateral sinus to the end of
the element.

Description. (P) element. The blade is straight to
slightly curved. The profile commonly has a high peak
anteriorly, is more or less flat at mid-blade and tapers to
a point posteriorly, but in some specimens, the anterior
profile is only slightly higher than it is at mid-length.

The lower profile may be nearly straight or may have the posterior half of the unit arched. The denticles are broad, usually discrete, although some fusion occurs in the middle and posterior parts of the blade. They are poorly differentiated below the crest of the blade. Commonly the anterior denticle is significantly larger than the others and is triangulate in shape. Denticle number ranges between 6 and 13.

The platform is strongly asymmetrical. The outline of the narrower lobe is variable with some specimens having a sulcus and some not. Its transverse profile is sloping or angulate (Text-fig. 6). The wider lobe is about 25 percent as long as the element, has a sloping or angulate profile and an anterior margin nearly at right angles to the blade. Non-platform elements. The elements are generally similar to those of A. sexidentata, but generally are larger. (A_1) develops a broader, more spatulate denticle at the higher stratigraphical levels.

Remarks. A. johnsoni alpha morphotype intergrades morphologically with both A. johnsoni beta morphotype and A. sexidentata. It is interpreted as the ancestor of the beta morphotypes whose stratigraphical ranges it precedes and overlaps.

Material: 309 (45).

beta morphotype

1969. Spathognathodus johnsoni Klapper, pl. 5, figs. 8-16.

1971. Spathognathodus johnsoni Klapper, in Klapper and others, pp. 289-290.

1973. Ozarkodina johnsoni (Klapper), Klapper, in Ziegler, ed., Ozarkodina pl. 2, fig. 1, pp. 235-236.

1977. Ozarkodina johnsoni (Klapper), Klapper, p. 52.

1980. Ozarkodina johnsoni (Klapper), beta morphotype Klapper and Murphy, p. 498, fig. 2, nos. 4, 15, 17-19.

Table 4. Occurrence data for the elements of the <u>Amydrotaxis</u> <u>sexidentata</u> n. sp. and <u>A</u>. <u>Johnsoni</u> beta apparatuses.

TABLE 4A

AMYDROTAXIS SEXIDENTATA n. sp.

	Footage	P	O	N	B_1	B_2	B_3	UCR #
BC II	713.5	1	–	–	–	–	–	7405
	720	2	–	–	–	–	–	935
	868	4	–	–	–	–	–	936
	871.5	1	–	–	–	–	–	8902
	875	–	–	–	–	1	–	8903
	883	–	–	–	–	–	1	8904
	1050	1	–	–	–	–	–	5458
	1086	1	–	–	–	–	–	937
	1100	2	–	–	–	–	–	8906
	1107	1	–	–	1	–	–	8907
	1685	6	1	1?	2	5	–	8908
SP I	1617	12	1	–	5	6	2	8977
	1658	4	1	–	4	2	–	8979
	1674	1	–	5	–	–	–	8980
	1665	–	–	–	–	–	–	8981
	1672	4	–	–	2	–	1	8982
	1733	5	–	–	2	1	1	8988
SP VII	150	–	1	–	–	–	1	8748
	176	11	4	–	6	7	1	8752
	181	9	8	–	7	5	1	7007
	200	1	–	–	–	1	–	8758
SP VIII	36	–	1	–	–	1?	–	8909
	103	1	–	–	–	–	–	8910
MC	17.6	4j	–	–	–	–	–	8517
	22.3	1	–	–	–	–	–	8519
	27.5	2	–	–	–	–	–	8523
	28	4	–	–	–	–	–	8524
	28.25	1	–	–	1	1	–	8525
COP V	105	1	–	–	1	–	–	6295
	110	1	–	–	1	–	–	6296
	115	2	–	–	–	–	–	7404

TABLE 4A - continued
AMYDROTAXIS SEXIDENTATA n. sp.

	Footage	P	O	N	B_1	B_2	B_3	UCR #
COP V	121	1	-	-	-	1	-	6297
	145	2	-	-	-	-	-	6298
COP IV	145	4	-	-	1	-	-	6237
	149.5	1	-	-	-	-	-	6239
	179.5	6	1	-	1	-	-	6238

TABLE 4B
AMYDROTAXIS JOHNSONI beta

		P	O	N	B_1	B_2	B_3	UCR #
SP VIII	303	3	-	-	-	2	-	8927
	307	-	-	-	1	1	-	8929
	308	3	-	-	-	-	-	8930
	311	1	-	-	-	-	-	8931
SS II	-	1	-	-	1	-	-	7400
	-	3	2	-	-	-	-	7401
II A	-	4	-	-	-	-	-	7402
	-	1	-	-	1	1	-	7403
COP IV	276	2	-	-	-	-	-	8888
	377	1	-	-	-	-	-	6265
	400	1	-	-	-	-	-	6267
	404	1	1	-	-	-	-	7407
	413	-	1	-	-	-	-	7409
	414	1	-	-	-	-	-	6268
IV A	410	-	1	-	1	-	-	6272
SP VII	391	1	3	-	-	1	-	8785
MC	104	2	-	-	-	-	-	8550
	107	2	-	-	-	2	-	8551
	109	1	-	-	-	-	-	8552
	121	1	-	-	-	-	-	8556
	134.5	2	-	-	-	-	-	8559
	144	2	-	-	-	1	-	8560
	160	1	-	-	-	-	1	8565

Diagnosis. O. johnsoni beta morphotype is characterized by the shouldered transverse profile co-occurring with the sulcate outline of the narrow lobe.

Description. (P) element. The blade is nearly straight with nearly flat upper and lower profiles except posteriorly where the upper profile descends to meet the lower in a point. Denticles are short, vary in number, are discrete or fused, and more or less even in size and spacing.

The platform is asymmetrical with the narrow lobe bilobate and the wider lobe with the anterior margin oriented nearly perpendicular to the blade. The platform typically has a shouldered transverse profile on both lobes. The element is relatively thick.

Non-platform elements are poorly known (table 1). (N) and (A$_3$) elements have not been identified and the others are known only from a few scraps. Undoubtedly, the johnsoni beta apparatus data constitute the weakest point in the reconstruction of the Amydrotaxis apparatus.

Remarks. A. johnsoni beta (P) element intergrades morphologically and descends from A. johnsoni alpha morphotype. Its certain descendant A. cf. A. johnsoni (Klapper, 1969, pl. 5, figs. 17-20) from Royal Creek Section 1 in Yukon Territory, Canada has not been found in Nevada.

Polygnathus Hinde, 1879

The earliest known form of the genus Polygnathus is P. pireneae Boersma. It has been considered as a derivative of Eognathodus and as the root stock of the genus (Klapper and Johnson, 1975), but more recently Lane and Ormiston (1979, p. 47) have suggested that it was derived from an ozarkodinid radical. No truly adult intermediate forms have yet been described that would favor any of the previous theses. However, from its stratigraphic position in the Salmontrout Formation (Lane and Ormiston, 1979, table 1) and in Nevada (table 3) and the lack of intermediate forms, it is certain that the ancestor is not a phyletic descendant of

Eognathodus and that it has to be in the sulcatus Zone or
earlier.   It must, therefore, be found in the early
Eognathodus morphs or, as suggested by the ontogenetic
studies of Lane and Ormiston (1979), in the late Ozarkodina
pandora morphs, since these are the forms with an open basal
cavity somewhat similar to that of P. pireneae.

Polygnathus pireneae Boersma, 1973
Pl. 1, Figs. 33-38

1973.   Polygnathus pireneae Boersma, p. 287, pl. 2, figs.
        1-12.

1972.   Polygnathus lenzi Klapper. Uyeno, in McGregor and
        Uyeno, 1972, pl. 5. figs. 10-12.

1977.   Polygnathus pireneae Boersma. Klapper, in Ziegler,
        ed., 1977, pp. 489-490.

1977a. Polygnathus boucoti Savage, 1977a, pl. 1, figs.
        13-28.

1977.   Polygnathus n. sp. R. Al Rawi, 1977, pl. 5, fig. 47.

1979.   Polygnathus pireneae Boersma. Lane and Ormiston, p.
        62, pl. 3, figs. 15-17, pl. 5, figs. 2, 3, 9, 10,
        27-34, 37.

1980.   Polygnathus pireneae Boersma. Klapper and Johnson,
        p. 454.

Remarks.   Eight specimens of Polygnathus pireneae have been
recovered recently from a very large sample (+40 kg) at COP
II 295.   Another sample nearby, UCR 7399, that is not in a
measured section, also yielded five specimens of the
species.   These specimens and those of Sandberg (1979, p.
90) from the Slaven Chert are the only known records of its
presence in Nevada.   The morphology of the Nevada specimens
compares favorably with previously illustrated specimens,
but most of those previously illustrated are small and not
well preserved.   The species is, thus, not well characterized and its variation largely a matter of speculation.

    Boersma (1973, p. 287) has described this species as
lacking adcarinal grooves.   His specimens are very small,
all less than-one third the size of the Nevada specimens,

and it is probable that larger (P) elements would show the development of adcarinal grooves. The larger of the Nevada specimens show distinct adcarinal grooves developed on the anterior part of the platform.

One of the more consistent features of the morphology is the ornament of the posterior platform. Only one of the presumably adult specimens (Lane and Ormiston, 1979, pl.5, fig. 2) shows the carina fused to the posterior end of the unit. In the others it consists of a series of nodes.

According to Boersma, only his larger specimens showed the posterior constriction of the basal cavity. This is in accord with the material from Alaska and Nevada.

We follow Klapper (in Klapper and Johnson, 1980, p. 454) and Lane and Ormiston (1979, p. 62) who synonymized Polygnathus boucoti Savage with P. pireneae, however, P. boucoti may be distinct. The specimens of P. pireneae of Boersma and of Lane and Ormiston are as small or smaller than P. boucoti, yet all distinctly show the lateral rows of nodes whereas P. boucoti shows them weakly in only one specimen. The carina in P. pireneae is approximately equally developed or less well developed than the lateral ornament. In P. boucoti the carina is prominent and the lateral ridges weak or absent. In either case, the specimens are very small and juveniles are notoriously difficult to identify.

Range. It is now obvious that P. pireneae is much younger than Boersma originally supposed (1973, p. 288). Its minimal range based on the well-controlled Alaskan sections of Lane and Ormiston (1979) and its occurrence ·at COP II in Nevada indicates a lower limit within the range of Pedavis mariannae. It occurs in Nevada with Pedavis mariannae and Eognathodus sulcatus mu morphotype at both localities. Lane and Ormiston show it ·beginning approximately in the middle of the range of their E. sulcatus kindlei and overlapping the lower range of P. dehiscens (1979, table 1).

Description of the Nevada specimens. All specimens show an inflated basal cavity visible in the upper view approximately in the center of the unit. The posterior platform is more or less pointed, with the outer margins ornamented by rows of short transverse ridges which, with one exception, terminate at a moderately developed groove which runs the length of the platform. The carina is fused in the anterior part of the platform, but is formed by a series of equant nodes posteriorly. The blade is short, less than one-third of the length. The platform is almost straight or curved in upper view and tapers gradually in both directions from its maximum width at mid-platform. Diagnosis. Needs to be modified.

FAMILY UNCERTAIN

Erika new genus

Type species. Erika divarica n. sp.
Derivation of the name. After Erika L. Erickson.
Diagnosis. A genus of conodont with six bar-like skeletal elements divided into anterior and posterior processes that bear divergent denticles along all or part of their lengths.
Remarks. The genus is known only from a single species and bears no close relationship to other Devonian genera. It resembles Oulodus Branson and Mehl as reconstructed by Sweet and Schönlaub (1975) in having no platform element and six bar-like elements. Also some of the element shapes show a resemblance to those of Oulodus.

The genus appears in the lower delta Zone and ranges up to the top of the delta Zone. At present only a small number of specimens is available, but the very distinctive nature of the apparatus warrants its description.

Erika divarica n. sp.
Pl. 6, Figs. 1-13

Syntypes. UCR 8776/2, 4; 8993/6; 8765/3, 12; 8766/3; 8767/5; 8771/4; 7410/1.

Type locality. SP VII section, east side Coal Canyon, northern Simpson Park Range, Eureka County, Nevada.

Diagnosis. See generic diagnosis.

Description. Six elements are inferred to constitute the apparatus. No platform element is known. Each of the elements bears short conical or keeled denticles arranged so that each alternate denticle points in a different direction. Because of the similarity to Oulodus mentioned above, they are arranged so that the similar elements will occupy the same relative position in the apparatus and the notation of Sweet and Schönlaub (1975) is used to facilitate the comparison with Oulodus.

Material:76 (15).

(Pa) element (Text-fig. 7a) (Pa) has unequal processes. The cusp is erect and offset laterally from the adjacent denticles. It is not possible from the curvature of the denticles to determine a conventional orientation, so the description of (Pa) will be in terms of the shorter or the longer process. The shorter process has only four denticles and has an open basal groove to its distal end. In the longer process the basal cavity has already narrowed to a groove at a position close to the cusp. The basal cavity has a lenticular shape, slightly expanded to a lip on one side.

(Pb) element (Text-fig. 7c) (Pb or M) is a twisted V-shaped element with a prominently protruding basal cavity which inverts as it extends distally under the processes. The cusp is only slightly larger than the neighboring denticles, but markedly divergent in its orientation. This element also shows similarity to Oulodus jeannae (M) element (Sweet and Schönlaub, 1975, pl. 1, fig. 18).

(Sa) element (Text-fig. 7d) (Sa) has unequal processes. The cusp is on the inner side of the element and curves gently toward the inner side. The basal cavity is a small asymmetrically lenticular depression surrounded on both sides by an inverted cavity which extends to the ends of

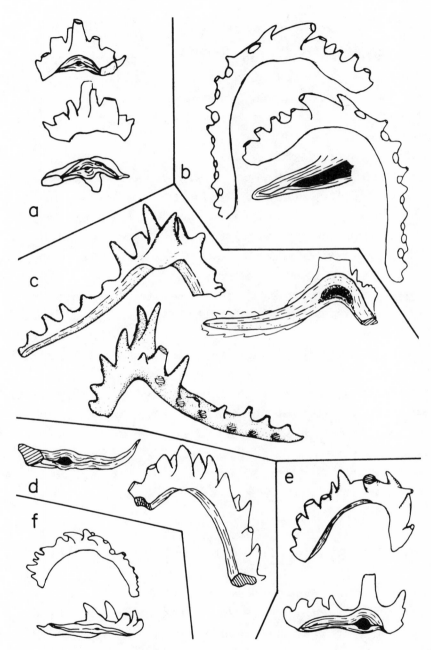

Text-figure 7.   Apparatus elements of <u>Erika</u> n.g.   Camera lucida drawings of: a.- Pa element, front, back and basal views, SP VII 348 feet; b.- Sc element, lateral and basal (showing inverted cavity) views, WP, Cortez Range; c.- Pb or M element front, back and basal views, basal shows small normal cavity surrounded by inverted cavity; SP VII 359 feet; d.- Sa element, side views, SP VII 274 feet, and basal view specimen from unknown locality; e.- Sb or Pb element, side and basal views, COP IV 164 feet; f.- Sb element, side and basal views, SP VII 250 feet.

both processes.   The cusp is distinctly larger than the other denticles.

(Sc) element (Text-fig. 7b) (Sc) resembles (Sa) in its broad U-shape, but the cusp is larger and the basal cavity extends part way along the processes before inverting. Also the anterior process is twice as long as the posterior process.

(Sb) element (Text-fig. 7f) (Sb) is a broad U-shaped element with an almost entirely inverted basal cavity and a laterally inclined cusp that is little different in size from the other denticles. The only part of the basal side of the element not inverted is a small depression below the cusp that varies from only about one-half the size of the width of the bar to most of the width of the bar. The two processes are approximately the same length with every other denticle curving laterally and the alternate ones being more erect.

Another element (Text-fig. 7e) is also U-shaped, but, unlike (Sa) and (Sc) elements, the processes do not lie in the same plane. The cusp is larger than the other denticles and is strongly curved laterally. The basal side of the element is inverted around a small depression and beneath the two processes. The anterior process is about two-thirds as long as the posterior process. This element could be an (Sb) or, possibly, a (Pb) element.

Family Icriodontidae Müller and Müller, 1957

Genus Pedavis Klapper and Philip, 1971
Klapper and Philip (1971, p. 446) proposed the genus Pedavis for what they believed to be three-element skeletal apparatuses in which the platform element has four processes and is shaped like a bird's foot. They classified all apparatuses of this type as Type 4 apparatuses and listed Pedavis, Icriodus, and Icriodella as examples. The notation for the elements of the apparatus given by Klapper and

Philip (1971, p. 438) is: (I) for the platform element; (S_1) for the pyramidal element (= saggitodontan element); (M_2) for the conical element. We follow Klapper and Philip in their notation of the (I) and (S_1) elements, but we differentiate ,the conical elements according to their morphotype as (M_{2a}), (M_{2b}), etc. (Text-fig. 8).

A principal problem in making a classification and elucidating a biostratigraphic scheme for these Pedavis species is their low abundance in virtually all collections from which they are described. The only taxon for which significant numbers of specimens are available is Pedavis pesavis. At present, we recognize the following specific or subspecific-rank taxa in Nevada: P. latialata, P. pesavis, Pedavis pesavis n. ssp. A Klapper and Philip, P. biexoramus n. sp. (= Pedavis sp. C of Klapper and Murphy), P. mariannae Lane and Ormiston (= P. n. sp. A of McGregor and Uyeno), P. brevicauda n. sp. (=Pedavis n. sp. ·A of Klapper), and P. breviramus n. sp.

An additional taxon from Nevada is mentioned and figured in Murphy and others, 1979, but remains to be described. This taxon is Early Silurian in age and significantly extends the range of the genus providing that the specimens are properly identified as Pedavis. This may still be regarded as doubtful because the other apparatus elements have not yet been identified. Klapper and Philip (1971, 1972) hypothesized that 3 elements are associated in the skeletal apparatus of Pedavis rather than the six suggested as the basic number in Ozarkodina and Pandorinellina of the Devonian or the three suggested from reconstruction of the Icriodus apparatus. The present work suggests that Pedavis has a set of transition elements that were all grouped as (M_2) elements by Klapper and Philip. The number of these is not well documented, but there are certainly three and there may be four of them (Text-fig. 8). These have already been figured by Klapper and Philip (1971, fig. 14; 1972, pl. 3, figs. 1-6, pl. 4, figs. 1-4, 15, 16, 18-21).

Pedavis

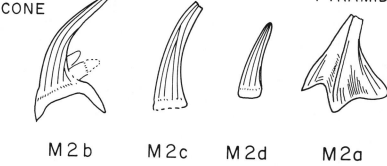

COMPLEX CONE   SIMPLE CONES   PYRAMID

M2b   M2c   M2d   M2a

Text-figure 8.   Conical elements of the *Pedavis* apparatus.

(M_{2a}) is pyramidal in shape like the (S_1) element, but with only a cusp and lacking denticles.

(M_{2b}) is a curved conical element with one to three small, posterior denticles.

(M_{2c}) is a simple cone with asymmetrical ornamentation and pronounced posterior curvature.

(M_{2d}) is a simple cone with asymmetrical ornamentation and only a slight-posterior curvature.

Recent work by Lane and Ormiston (1979) in Alaska, and by Uyeno (1981) in the Canadian Arctic, demonstrates that many more discoveries need to be made before the details of Pedavis evolution will be understood. At present, the lack of variation studies on large samples from single horizons of any part of the apparatus inhibits the stratigraphic use of all Pedavis species. In addition, assessment of the variability of P. pesavis is clouded by the apparently mixed fauna from the holotype locality (Bischoff and Sannemann, 1958).

The known Pedavis morphotypes and their stratigraphic sequence suggest that the basic plan of the platform elements of the genus is represented by P. latialata-P. pesavis which are close morphologically. Starting with P. latialata, which is characterized by a sigmoidal axis constituted by the main and posterior processes with short inner and outer lateral processes. In Nevada the stratigraphic sequence shows this basic form is present in the latialata Zone, reappears again slightly altered in the delta Zone (pl. 7, figs. 4, 5) and at the base of the kindlei Zone (pl. 8, fig. 3). Fig. 9 is based on the assumption of continuity of this stock in the intervening strata. Uyeno's (1981, pl. 5, figs. 8, 9) specimen from the 'Delorme' Formation, District of Mackenzie is morphologically intermediate and occurs in the hesperius Zone, thus supporting this assumption. The description of the new species P. biexoramus and its comparison with various Spanish species (Carls, 1969, 1975; Carls and Gandl, 1969) drew our attention to the lack of definition of the parts of the platform element of Pedavis, particularly the separation of the anterior and posterior processes. In Icriodus, the posterior process originates at the posterior peak of the double-tipped basal cavity. Since P. biexoramus also seems to have anterior and posterior tips of the basal cavity, we have selected the posterior tip of the basal cavity as the point separating the main and posterior processes. In most

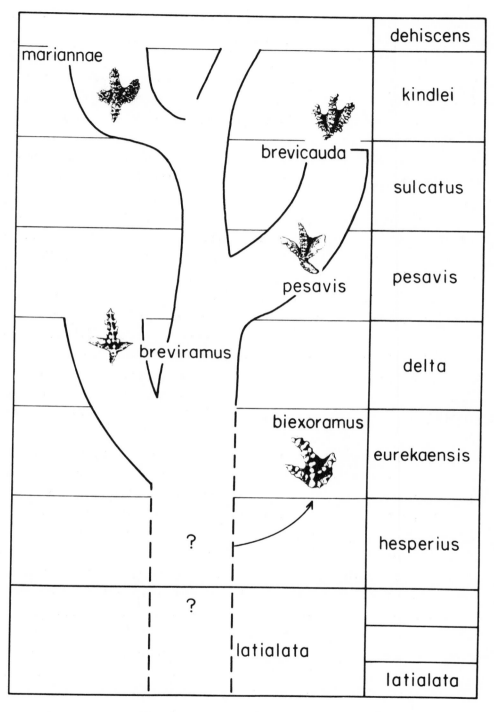

Text-figure 9.   Ranges and hypothesized relationships of species within the Genus Pedavis.

Pedavis species we have not seen the double tip that is present in *Icriodus*.

Pedavis pesavis (Bischoff and Sannemann, 1958)
(I) element
Pl. 7, Figs. 13, 20

1958. *Icriodus pesavis* Bischoff and Sannemann, pp. 96-97, pl. 12, figs. 1, 4, 6, 7 (only) [fig. 1 = holotype].

1969. *Icriodus pesavis* Bischoff and Sannemann. Klapper, pp. 8-10, pl. 1, figs. 1-3, 5-11, 13, 14.

Not 1966. *Icriodus pesavis* Bischoff and Sannemann. Clark and Ethington, p. 680, pl. 83, fig. 5 [= broken (I) element of *Icriodus* sp. indet.].

1971. *Pedavis pesavis* (Bischoff and Sannemann). Klapper and Philip, pp. 446-447, Text-fig. 14 (I), (S_1), and (M_2) elements.

1972. *Pedavis pesavis pesavis* (Bischoff and Sannemann). Klapper and Philip, pp. 102-103, pl. 4, fig. 10 (I) element.

Remarks. Nevada specimens of *P. pesavis* (I) are similar to those specimens previously reported from Europe, the Canadian Yukon, and central Nevada. *P. pesavis* is restricted to the *pesavis* Zone according to Klapper (1977, p. 38).

According to Klapper and Philip (1972, p. 103) the (I) element in *pesavis* Zone has the long, sigmoidal posterior process which contrasts with the shorter, straighter posterior process of *P. breviramus* n. sp. in the underlying *delta* Zone.

Material: 53 (15).

Pedavis sp.

1968. *Icriodus pesavis* Bischoff and Sannemann. ? Schulze, p. 191, pl. 16, fig. 8 (only).

1969. *Icriodus pesavis* Bischoff and Sannemann. Carls, 1969, pp. 330-332, pl. 1, figs. 2, 3.

1975. *Icriodus pesavis* Bischoff and Sannemann. Telford,
        pl. 4, figs. 2, 3 (only).

Diagnosis. A species of *Pedavis* in which the (I) element is characterized by a K-shaped basal cavity and has a nearly straight main process.

Description. The (I) element is pedaviform with unequally developed processes that give the unit an asymmetrical appearance. The main process is nearly straight, ornamented by three rows of denticles or by two outer rows of denticles connected to a central longitudinal ridge by transverse ridges. The lateral processes are also straight, the outer is more anterior than the inner (Carls, 1969, pl. 1, figs. 2, 3). They maintain angles of about 45 degrees with the main process and have irregular transversely arranged denticles or ridges as ornament. The basal cavity is open and without folded-over margins as is characteristic of *P. pesavis* and *P. latialata*.

The posterior process is short and extends almost in line from the inner lateral process.

Remarks. Carl's specimens occur about 25 m above *O. transitans* which has a known range limited to the *delta* Zone and *pesavis* Zone (Klapper, 1977, p. 38). Thus the position of the taxon is in the *delta* Zone or higher.

Pedavis brevicauda n. sp.
Pl. 6, Figs. 14-17

1969. *Icriodus* n. sp. A Klapper, p. 10, pl. 1, figs. 15-18.
1971. *Icriodus* n. sp. A Klapper. Klapper and others, p.
        290, Text-fig. 1 (I element).
1971. (not) *Icriodus* sp. A Klapper. Fåhraeus, p. 675,
        pl. 78, figs. 1, 2 (I element).

Holotype. UCR 6211/1. See also synonymy of Klapper and Johnson, 1980, p. 451.

Derivation of the name. Brev + caud, Latin = short tail.

Diagnosis. A species of *Pedavis* in which the (I) element is characterized by the short, peg-like posterior process that is either straight or sharply deflected laterally and, in

lower view, by the circular outline of the posterior margin
of the basal cavity.

Upper view. The general shape of the element is pedaviform.
The unit possesses a main process, two well-developed
lateral processes, and a peg-like posterior process. The
relative lengths of the lateral processes are variable. The
three anteriorly directed processes have longitudinal axes
that are straight or only slightly curved. Three types of
denticulation patterns occur on the main and lateral
processes. On most specimens the anterior processes bear
two longitudinal marginal rows of denticles. In the second
type there is a median longitudinal trough-like depression
between the marginal denticle rows that is occupied by a row
of very tiny denticles. The third type has a bar-like
connecting ridge between the outer denticles. Only one very
large specimen exhibits a partially developed central
longitudinal ridge (pl. 6 fig. 16). Some (I) elements have
processes that bear an irregular ornamentation pattern.

   The posterior process is a short peg-like projection
that either extends posteriorly or is deflected strongly
toward one or the other lateral margin. The projection may
bear one to three prominent costae or striae.

Pedavis biexoramus n. sp.
Pl. 5, Figs. 28, 33

1975. Pedavis sp. nov. C Klapper and Murphy, (1974), p. 50,
      pl. 12, fig. 12 (I element).
See also synonymy of Klapper and Johnson, 1980, p. 451.

Holotype and Derivation of the name. UCR specimen 9035/1;
bi + exo + ram, Latin = two outside branches.

Diagnosis. A species of Pedavis in which the (I) element is
characterized by the strong lateral deflection of the
posterior process.

Description. The general shape of the (I) element is
pedaviform. It possesses a main process and well-developed
inner and lateral processes. All are directed laterally or

anteriorly.   The posterior process is well developed and directed laterally or slightly anteriorly.

The longitudinal axis of the main process is straight to moderately curved.   It bears transverse ridges that usually show three weakly developed denticles.   Deep grooves separate the transverse ridges and denticle sets.   A thin, weakly developed median longitudinal ridge is variably present in the grooves.

The longitudinal median ridge extends posteriorly beyond the last denticle-set on the main process and connects three isolated denticles.   These denticles are round, and are about the same height as the denticle-sets on the main process.

The inner lateral process is well developed.   In large specimens it bears two to four transverse ridges that connect faintly developed marginal denticles; round discrete denticle-sets are rarely developed.   On smaller specimens the transverse ridges have been compressed laterally to produce a single row of round denticles or carina linked by a longitudinal ridge (e.g., Klapper and Murphy, 1975, pl. 12, fig. 12).   The longitudinal axis of the process is straight, or is faintly curved with the concave side facing anteriorly.   The inner lateral process joins the main process at the middle isolate cusp of the three-cusp series.

The outer lateral process is weakly developed (to absent?).   In those specimens that clearly possess an outer lateral process it consists of a weak ridge that bears two to five denticles or a carina.   The longitudinal axis of the process is faintly curved, with the concave side facing anteriorly.   The process joins the main process slightly forward of the middle isolate denticle of the three-denticle series.   The specimen illustrated by Klapper and Murphy (1975, pl. 12, fig. 12) has an outer lateral process that either lacks denticles entirely (thus suggesting an ontogenetically earlier stage), or has been broken such that the denticulated distal end of the process is missing.   At Copenhagen Canyon the outer lateral process has clearly been broken off of some (I) elements. However, with some specimens that are missing an outer lateral process there is no clear evidence of fracture along the edge of the element. These specimens may be (I) elements that represent an early ontogenetic stage in which an outer lateral process was not yet developed. Such forms have an embayed margin on the outer lateral side of the element, a feature hardly noticeable in larger specimens. The specimen illustrated by Klapper and Murphy may represent an intermediate ontogenetic stage in which a non-denticulated outer lateral process is weakly developed and in which the platform embayment is partially closed.

The posterior process is moderately to well developed. It consists of a weak ridge that bears two to four round denticles or a carina. The longitudinal axis of the process is staight, or faintly curved with the concave side facing anteriorly. The posterior process is sharply deflected toward the outer lateral margin of the unit. The process joins the unit at the last isolate denticle of the three-denticle series. From this denticle, the axis of the process extends straight toward the outer margin of the unit, then curves faintly in an anterolateral direction. The juncture between the posterior and main processes forms a 90-100° angle.

The basal cavity is open. The posterior margin of the element is smoothly rounded. The anterior margin is embayed between the main and inner lateral processes. The in-folded portion of the posterior margin of the outer lateral process that is present in the (I) elements of *P. pesavis*, *P. latialata*, and *P. brevicauda* does not occur in *P. biexoramus*.

Remarks. Two of the fragmental elements of an *Icriodus* figured by Graves (1952, pl. 81, figs. 14, 16) superficially resemble *P. biexoramus*, but the angle of divergence of the processes and style of ornamentation seem different.

In P. pesavis (I) and other pedaviform (I) elements the
posterior process is separated from the main part of the
unit by a deep sinus or embayment in the outer lateral
margin of the platform. By contrast, in P. biexoramus (I)
the posterior process is connected to the outer lateral
process by a smooth expanse of platform surface that bridges
the gap. Hence, the deep embayment does not occur on the
outer side of the unit.
Material: 12 (6).

Pedavis mariannae Lane and Ormiston, 1979
Pl. 8, Figs. 1, 2, 4-8, 10
?1971. Icriodus n. sp. A Uyeno, in McGregor and Uyeno, pl.
5, figs. 36-38.
1979. Pedavis mariannae Lane and Ormiston, pp. 59-60, pl.
4, figs. 14-20, 23-25, 27; pl. 5, figs. 1, 7, 11-14,
17-22.
Diagnosis. A species of Pedavis in which the (I) element is
characterized by a posterior process that extends straight
back from the junction of the lateral processes and an adult
ornamentation pattern of branching ridges on all processes.
Description.   The   general   shape   of   the   element   is
pedaviform.   The   outer   lateral   process   is   equal   to   or
shorter in length than the main process and directed at an
angle of about 45° to the main process. The inner lateral
process   is   about   half   the   length   of   the   outer   lateral
process and is directed at an angle between 45° and 90° to
the main process. The anterior and posterior processes are
connected by a straight, thin median ridge that passes
through the junction point of all the processes. All
processes are relatively massive, and flat topped in
comparison to other Pedavis and in mature specimens they are
ornamented by a series of more or less interconnecting
ridges as opposed to discreet denticles.
The longitudinal median ridge of the main process
extends posteriorly and forms the longitudinal axis for the

posterior process.   Two types of posterior process are
developed: (1) a broad, robust, and relatively short type
(pl. 8, fig. 1), and (2) a relatively long type (pl. 8, fig.
6).   The first type occurs stratigraphically lower in the
Copenhagen Canyon area than the second type.   The two types
are designated alpha and beta in all figures and tables.
(S_1) elements like those found by Lane and Ormiston (1979,
pl. 5, figs. 20-22) have not been found in Nevada.   In the
present collections the only elements approaching the style
of the (M_2) elements are a few unornamented cones similar
to Acodus.

Pedavis breviramus n. sp.
Pl. 7, Figs. 1-6, 12

1968.   Icriodus pesavis Bischoff and Sannemann.   Schulze,
        pl. 16, fig. 5 (only).
1969.   Icriodus pesavis Bischoff and Sannemann.   Klapper,
        pl. 1, fig. 12 and ? fig. 4.
1970.   Icriodus pesavis Bischoff and Sannemann.   Seddon, pp.
        55-56, pl. 4, figs. 21, 23.
1972.   Pedavis pesavis Bischoff and Sannemann.   Klapper and
        Philip, pl. 3, figs. ?4-9 only.
1976.   Pedavis pesavis (Bischoff and Sannemann) Savage,
        1976, pl. 2, figs. 10-17.
1981.   Pedavis n. sp. A. Uyeno, pl. 7, figs. 41-48.

Holotype.   UCR 8767/9, SP VII 274 feet.

Derivation of the name.   brev + ram, Latin = short tail.

Diagnosis.   A species of Pedavis in which the (I) element is
characterized by two short lateral processes at high angles
to the main process, a short slightly sigmoidal, posteriorly
directed posterior process, and four or five transverse
denticle rows on the main process.

Description.   The (I) element is pedaviform with a slightly
arched main process, two short lateral processes at high
angles to the main process, and a posterior process that is
slightly sigmoidal.   The main process is large with four or
five transverse denticle rows connected by a prominent

longitudinal ridge. The outer lateral process is at a high angle (up to 90°) to the main process, sub-equal in width and ornamented with a series of radiating filae. The inner process is much smaller and ornamented by a set of transverse denticles or ridges connected by a thin longitudinal ridge. The posterior process is similar to it in size, shape, and ornament.

Material: 20 (12).

Remarks. P. breviramus n. sp. is separated from P. pesavis subsp. A. Klapper and Philip (1972, p. 103) on the basis of the angles at which the processes intersect. In P. breviramus the lateral processes diverge from the inter- section with the posterior process at a high angle. In P. pesavis subsp. A, as restricted here, the posterior and one lateral process are in line with one another. See especially Carls, 1969, pl. 1, figs. 2 and 3.

P. breviramus n. sp. occurs in the Devon Island Formation in the Canadian Arctic in association with Icriodus woschmidti hesperius and Notoparmella gilli (Uyeno, 1981, fig. 23), species that are found in the eurekaensis Zone in the lower Windmill Limestone in Nevada.

Genus Icriodus Branson and Mehl, 1938

The taxonomy of North American Icriodus has been reviewed recently by Klapper (in Klapper and Johnson, 1980). He recognizes five taxa that occur in the lower half of the Lower Devonian in central Nevada. Two additional taxa are described here, but are not formally named.

The earliest Icriodus species in Nevada, Icriodus woschmidti hesperius Klapper and Murphy ranges into the lowermost part of the Windmill Limestone. Aside from these occurrences, however, the lower and middle parts of the Windmill are almost devoid of Icriodus. A few rare specimens of Icriodus sp. and Icriodus n. sp. G (Klapper, 1977) are present, but these representatives of the genus do not seem to be closely related to the earlier or later

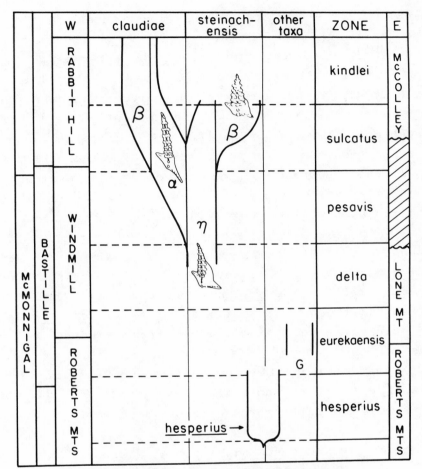

HYPOTHESIS OF RELATIONSHIP AND
STRATIGRAPHIC DISTRIBUTION OF
ICRIODUS SPECIES

Text-figure 10.   Hypothesis of relationship and
stratigraphic distribution of Icriodus species in
central Nevada.   W indicate more western facies; E
indicates more eastern facies.

species in the section. The appearance of Icriodus steinachensis Al Rawi in the middle delta Zone marks the beginning of relatively abundant Icriodus faunas throughout the limestone belt of central Nevada. From this level through to the first occurrence of Polygnathus dehiscens we have been able to follow the evolution of the I. steinachensis lineage in detail. The important steps in this succession are shown in Text-fig 10. Essentially these are: 1) the appearance of the eta morph of I. steinachensis with a relatively low range of variability; 2) the increase in the range of variation to include, first, the alpha morph of I. claudiae and, then, the beta morph of I. steinachensis. The full range of variation between I. claudiae alpha and I. steinachensis beta persists through the middle part of the sulcatus Zone, but a morphologic gap is present between I. claudiae and I. steinachensis in the upper sulcatus Zone. At the base of the kindlei Zone both morphs of steinachensis disappear in our sections leaving only the variants of claudiae. The highest interval in the Rabbit Hill Limestone shows a predominance of claudiae beta, but a few of the alpha morphs are still present.

Icriodus steinachensis Al Rawi, 1977
Pl. 5, Figs. 31, 36

See Klapper and Johnson (1980) for synonymy.

Remarks. Klapper and Johnson (1980) have separated Al Rawi's (1977) species into two morphotypes, beta and eta, which correspond to Klapper's (1977) earlier Icriodus n. sp. B and Icriodus n. sp. H, respectively.

The morphotypes are distinguished on the basis of the outline of the spindle: morph beta has a carrot-shaped outline; morph eta has a spindle-shaped outline.

In this research we have looked at spindle shape, angle between the main process and the axis of the platform, the position of the point of junction of the processes, denticle

shapes and patterns, and size of the basal cavity. The
shape criteria seem to be the most reliable in terms of
stratigraphical use. In combination, shape and a few other
characters may eventually produce a finer subdivision.

Icriodus steinachensis eta appears in Nevada sections in
the delta Zone. The variation is relatively slight. The
specimens have the typical platform shape illustrated by
Klapper and Johnson (1980, pl. 2, figs. 25, 26) in which the
platform outline is lenticular and the platform axis is
arched (pl. 5, fig. 36, herein). The third or fourth
denticle row is the widest in the typical form and the angle
of the posterior lateral process ranges from less than 90
degrees to 110 degrees. Some specimens have a sharp raised
hair-like ridge on the spur and some do not. At maximum
width, the basal cavity is two to two and a-half times the
maximum width of the transverse denticle rows. This
ordinarily easily distinguishes I. steinachensis eta from
the later very similar I. n. sp. C of D. B. Johnson (Klapper
and Johnson, 1980, pl. 2, figs. 4-6= I. nevadensis Johnson
and Klapper, 1981). The lateral denticles are typically
almost circular in upper view, but there is extreme
variation in this respect from three distinct, rounded
denticles to a continuous transverse ridge. The mid-row
denticles vary from being sub-equal in size to the lateral
denticles and almost circular to being only slight swellings
where the longitudinal and lateral connecting ridges
intersect. The sizes of the cusp and the next anterior
denticle are variable.

The holotype of I. steinachensis Al Rawi is an eta
morph. It and 26 other eta morphs in Al Rawi's collection
at the Senckenberg Museum differ from Nevada specimens in
having a less well-developed spur and in the outer margins
of the basal cavity which are more parallel to the axis of
the main process than in the Nevada specimens.

Icriodus claudiae Klapper, 1980
Pl. 5, Figs. 32, 37

1980.   *Icriodus* aff. *I. celtibericus* Carls and Gandl. Klapper, *in* Klapper and Johnson, pl. 2, figs. 18, 23, 24.

1980.   *Icriodus claudiae* Klapper, *in* Johnson, Klapper, and Trojan, pp. 99-100, pl. 3, figs. 1, 9, 10.

Remarks.   Klapper has pointed out that the widely spaced anterior transverse rows of denticles and the closely spaced posterior ones consistently separate this species from the Spanish species *I. celtibericus* Carls and Gandl, which has evenly spaced rows throughout. In fact, this characteristic of Nevada forms generally separates them from European species other than *I. steinachensis*.

I. *claudiae* arises from *I. steinachensis* in the lower sulcatus Zone. The distinction between the two taxa is based on the possession of a raised central longitudinal ridge and an elongate spindle shape in the earliest *I. claudiae* (alpha morph) (pl. 5, fig. 37). The tendency in this branch of the lineage is for the spindle and basal cavity to become narrower, the spindle to straighten, and for an increase in the angle of the outer lateral process with the main axis of the spindle.

Two morphs are distinguished in *I. claudiae* on the basis of the size of the basal cavity relative to the width of the spindle. The earlier appearing alpha morph has the same type of expanded cavity as *I. steinachensis* eta. The cavity in *claudiae* alpha is two to three times the maximum spindle width. The beta *claudiae* morph has a basal cavity width less than two times the maximum spindle width (pl. 5, fig. 32).

The angle of the posterior lateral process to that of the spindle has no stratigraphic restriction within the studied interval. The posterior transverse denticle rows in some alpha morphs show a markedly raised central longitudinal ridge and a loss of the rounded lateral termination of the denticle rows.

One, two, or three single denticles or cusps may be present behind the transverse denticle rows. A small raised thread-like rib may be present on the spur.

An important problem in this group of species is the tendency for breakage to occur between the spindle and the basal cavity so that most specimens consist of spindles only, which are difficult to identify.

Icriodus woschmidti hesperius Klapper and Murphy, 1975

1969.   Icriodus woschmidti Ziegler. Klapper, p. 10, pl. 2, figs. 1, 2, only.

1975.   Icriodus woschmidti hesperius Klapper and Murphy, p. 48, pl. 11, figs. 1-19.

1980.   Icriodus woschmidti hesperius Klapper and Murphy. Klapper and Johnson, p. 449, pl. 2, fig. 11.

Remarks. Klapper and Murphy (1975, fig. 2), Klapper (1977, fig. 3), and Klapper (in Klapper and Ziegler, 1979, fig. 1) show the range confined to the hesperius Zone. We can now show that the range of I. woschmidti hesperius overlaps the range of Ozarkodina eurekaensis Klapper and Murphy in three sections (SP I, COP V, and BC II) and, hence, extends into the lower eurekaensis Zone.

No changes in the characterization of the taxon are required by our new data.

Icriodus n. sp. G of Klapper, 1977
Pl. 5, Figs. 29, 34

1977.   Icriodus n. sp. G of Klapper, p. 52, fig. 3.

1979.   Icriodus sp. Lane and Ormiston, pl. 1, fig. 29.

1980.   Icriodus n. sp. G of Klapper. Klapper and Johnson, pl. 2, figs. 1-3.

Remarks. This distinctive Icriodus is found rarely in the eurekaensis Zone. As originally noted (Klapper, 1977, fig. 3) the species is found at Coal Canyon and at Ikes Canyon. Our collections produced a few additional specimens all from

the _eurekaensis_ Zone.   The material is, thus, not yet
adequate to warrant formal taxonomic status.

Icriodus sp.
Pl. 5, Figs. 30, 35

A single specimen with a short platform relative to the
size of the basal cavity expansion was found in section SP I
at 1617 feet at Coal Canyon.   An almost identical specimen
has also been found at Ikes Canyon in a similar stratigraphic position.   The specimens are remarkably similar to
the holotype of _Icriodus_ _woschmidti_ _transiens_ Carls and
Gandl (pl. 15, fig. 1) from Spain, except for the progres-
sive widening of the spaces between the transverse denticle
rows toward the anterior that is seen in the Nevada
specimens.   One Spanish specimen (Carls and Gandl, 1969, pl.
15, fig. 3) also shows this tendency, but it is much less
developed in the Spanish material.   We do not synonymize _I._
woschmidti _transiens_ with _I._ _postwoschmidti_ Mashkova as has
been done by Klapper (in Klapper and Johnson, 1980, p. 448).

Bibliography

BERRY, W. B. N. AND M. A. MURPHY
    1975. Silurian and Devonian Graptolites of Central
        Nevada. Univ. California Publ. Geol. Sci.
        110:1-109.

BISCHOFF, G. AND G. SANNEMANN
    1958. Unterdevonische Conodonten aus dem Frankenwald.
        Notizbl. hess. Landesamt Bodenforsch.,
        86:87-110.

BOERSMA, K. T.
    1973. Description of certain Lower Devonian Platform
        Conodonts of the Spanish Central Pyrenees Leidse
        Geol. Med. 49:285-301, pls. 1-3.

BULTYNCK, P.
    1971. Le Silurien supérieur et le Dévonien inférieur
        de la Sierra de Guadarrama (Espagne Centrale).
        Deuxième partie: Assemblages de Conodontes
        a *Spathognathodus*. Bull. Inst. Roy. Sci. Nat.
        Belg. 47:1-43.

CARLS, P.
    1969. Die Conodonten des tieferen Unter-Devons der
        Guadarrama (Mittel-Spanien) und die Stellung des
        Grenzbereiches Lochkovium/Pragium nach der
        rheinischen Gliederung. Senckenbergiana
        lethaea, 50:303-355.

CARLS, P.
    1975. Zusätzliche Conodonten-Funde aus dem tieferen
        Unter-Devon Keltiberiens (Spanien).
        Senckenbergiana lethaea 56:399-428.

CARLS, P. AND J. GANDL
    1969.   Stratigraphie und Conodonten des Unter-Devons
            der Ostlichen Iberischen Ketten (NE-Spanien).
            N. Jb. Geol. Palaont. Abh., 132:155-218.

CHLUPÁČ, I., H. JAEGER, AND J. ZIKMUNDOVA
    1972.   The Silurian/Devonian boundary in the Barrandian.
            Bull. Canadian Petroleum Geol. 20:104-174

KAY, M. AND J. P. CRAWFORD
    1964.   Paleozoic facies from the miogeosyncline to the
            eugeosynclinal belt in thrust slices, central
            Nevada. Geol. Soc. Amer. Bull. 75: 425-454.

JOHNSON, D. B. AND G. KLAPPER
    1981.   New early Devonian conodont species of central
            Nevada J. Paleontol. 55:1236-1250.

KLAPPER, G. WITH A. R. ORMISTON
    1969.   Lower Devonian conodont sequence, Royal Creek,
            Yukon Territory, and Devon Island, Canada. J.
            Paleontol., 43:1-27, 6 pls.

KLAPPER, G. WITH CONTRIBUTIONS BY D. B. JOHNSON
    1977.   Lower and Middle Devonian conodont sequence in
            central Nevada in Murphy, M. A., W. B. N. Berry
            and C. A. Sandberg, eds., Western North America:
            Devonian, Univ. Calif. Riverside, Campus Mus.
            Contribs. 4:33-54.

KLAPPER, G. AND J. G. JOHNSON
    1980.   Endemism and dispersal of Devonian conodonts.
            J. Paleontol. 54:400-455.

KLAPPER, G. AND M. A. MURPHY
    1975.   Silurian-Lower Devonian conodont sequence in the
            Roberts Mountains Formation of central Nevada.
            Univ. Calif. Publ. Geol. Sci. 111: 62 pp.
            [imprint 1974]

KLAPPER, G. AND M. A. MURPHY
    1980.   Conodont zonal species from the delta and pesavis
            Zones (Lower Devonian) in central Nevada. Neues
            Jb. Geol. Paläont. Mh., 1980 (8):490-504.

KLAPPER, G. AND G. M. PHILIP

    1971. Devonian conodont apparatuses and their vicarious
        skeletal elements. Lethaia, 4:429-452.

KLAPPER, G. AND G. M. PHILIP

    1972. Familial classification of reconstructed Devonian
        conodont apparatuses. Geol. et Palaeontol. SB 1,
        97-114, Marburg.

KLAPPER, G., C. A. SANDBERG, C. COLLINSON, J. W. HUDDLE,
    R. W. ORR, L. V. RICKARD, D. SCHUMACHER, G. SEDDON, AND
    T. T. UYENO

    1971. North American Devonian conodont biostratigraphy.
        Geol. Soc. Amer. Mem. 127:285-316. Boulder,
        Colorado.

LANE, H. R. AND A. R. ORMISTON

    1979. Siluro-Devonian biostratigraphy of the Salmon-
        trout River area, east-central Alaska. Geol.
        et Palaeontol. 13:39-96.

MASHKOVA, T. V.

    1972. *Ozarkodina steinhornensis* (Ziegler) apparatus,
        its conodonts and biozone. Geol. et Palaeont.
        SB 1, p. 81-90, 2 pls.

MATTI, J. C., M. A. MURPHY AND S. C. FINNEY

    1975. Silurian and Lower Devonian basin and basin-
        slope limestones, Copenhagen Canyon, Nevada.
        Geol. Soc. Amer. Spec. Pap. 159, 48 pp.

MURPHY, M. A., J. C. MATTI, AND O. H. WALLISER

    1981. Biostratigraphy and Evolution of the *Ozarkodina*
        remscheidensis-Eognathodus *sulcatus* Lineage
        (Lower Devonian) in Germany and central Nevada.
        J. Paleontol. 55:747-772.

PICKETT, J.

    1980. Conodont assemblages from the Cobar Supergroup
        (Early Devonian), New South Wales. Alcheringa
        1980:67-88.

SANDBERG, C. A.
    1979. Devonian and Lower Mississippian Conodont
           Zonation of the Great Basin and Rocky Mountains,
           in Sandberg, C. A. and Clark, D. L., eds.,
           Conodont biostratigraphy of the Great Basin and
           Rocky Mountains: Brigham Young Univ., Geology
           Studies 26(3):87-106.
SAVAGE, N. M.
    1976. Lower Devonian (Gedinnian) conodonts from the
           Grouse Creek area, Klamath Mountains, northern
           California. J. Paleontol. 50:1180-1190.
SAVAGE, N. M.
    1977a. Lower Devonian conodonts from the Gazelle
           Formation, Klamath Mountains, northern
           California. J. Paleontol. 51(1):57-62.
SCHULZE, R.
    1968. Die Conodonten aus dem Palaozoikum der
           mittleren Karawanken (Seeberggebiet). N. Jb.
           Geol.-Palaont. Abh., 130:133-245.
UYENO, T. T.
    1981. Stratigraphy and Conodonts of Upper Silurian and
           Lower Devonian Rocks in the Environs of the
           Boothia Uplift, Canadian Arctic Archipelago, Part
           II. Systematic Study of Conodonts. Geol.
           Survey Canada, Bull. 292: 39-75.

PLATES

Plate 1--all figures approximately X30.

Figs. 1, 2. Ozarkodina excavata (Branson and Mehl) (P). 1, 2, lateral and top views of UCR 8765/13 from SP VII 250.

Figs. 3-9. Ozarkodina excavata tuma n. subsp., (P). 3, 4, lateral and top views of UCR 8536/2; 5, lateral view of UCR 8536/5; 6, lateral view of UCR 8536/8; 7, 8, top, basal, and lateral views of UCR 8536/1 from MC 55.75 feet.

Figs. 10-24. Ozarkodina pandora pi morph new morph (P). All elements from MC 160, Mill Canyon, Toquima Range. 10-12, basal, lateral and top views of UCR 8565/5; 13-15, top, lateral and basal views of UCR 8565/1; 16, 24, top and side views UCR 8565/4; 17-19, lateral, top and basal views of UCR 8565/3; 20, 21, top and side views of UCR 8565/2; 22, 23, top and side views of UCR 8565/6.

Figs. 25-32, 39, 40. Ozarkodina paucidentata n. sp. (P). 25-27, side, top and basal views, holotype, of UCR 7007/1 from SPVII 181, Coal Canyon, Simpson Park Range; 28, 29, top and side views of UCR 7007/11 from SPVII 181; 30-32, basal, top and side views of UCR 7007/2 from SPVII 181; 39, 40, top and side views of UCR 8752/1 from SPVII 176.

Figs. 33-38. Polygnathus pireneae Boersma (P). COP II 295, Copenhagen Canyon, Monitor Range. 33-35, top, basal and side views of UCR 8973/3; 36, 37, top and lateral views of UCR 8973/1; 38, top view of UCR 8973/5.

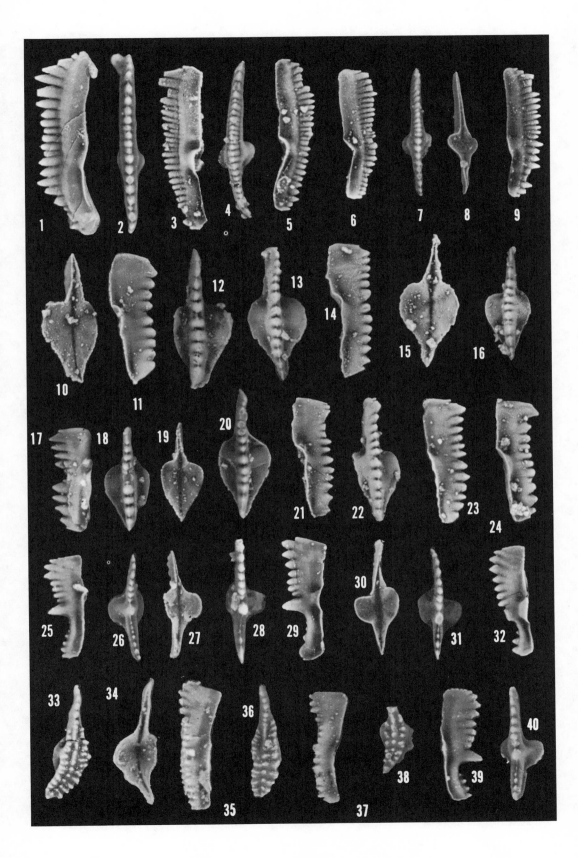

Plate 2--All figures approximately 30X.

Figs. 1-3. Pandorinellina cf. P. optima (Moskalenko) (P).
    Top, side and basal views of UCR 6299/19 from COP IV
    404.

Figs. 4-6. Ozarkodina remscheidensis (Ziegler) (P).
    Lateral views of UCR 8765/9, 10, 11 from SP VII 250.

Figs. 7, 8. Ancryodelloides asymmetricus (Bischoff and
    Sannemann) (P). Top and basal views of UCR 8776/5 and
    8776/18, respectively, from SP VII 359.

Figs. 9-11. Ancryodelloides transitans (Bischoff and
    Sannemann) (P). Side, top and basal views of UCR 8993/4
    from SP VII 348.

Figs. 12, 13, 15-17. Ancryodelloides omus n. sp.-A.
    transitans (Branson and Mehl) (P) intermediates. 12,
    16, top and side views of UCR 8767/8 from SP VII 274;
    13, 15 and 17, top side, and basal views of UCR 8768/1
    from SP VII 293.

Figs. 14, 18-29. Ancryodelloides omus n. sp. (P). 14, beta
    morph, top view of UCR 8920/1 from MC 81.5; 18-20, alpha
    morph, top, side and basal views of UCR 6247/1 from COP
    IV 120; 21-23, holotype, beta morph, side basal, and top
    views of UCR 6249/6 from COP IV 130; 24, 25, beta morph,
    top and side views of 6249/9 from COP IV 130; 26-29,
    beta morph, top views of UCR 8997/6 and 8997/4, and
    lateral and top views of UCR 8997/2 from the U Topolů
    section, Radotín Valley, Czechoslovakia.

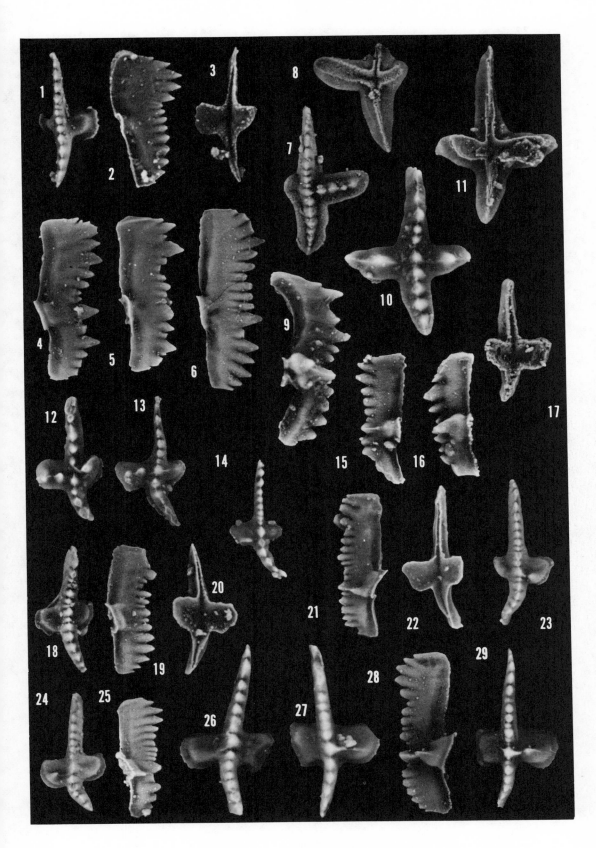

Plate 3--All figures approximately 30X.

Figs. 1, 2, 7, 8. Ozarkodina stygia (Flajs) (P), top and
    side views of UCR 8530/ 1 and 8530/2, respectively, from
    MC 3C.

Figs. 3-6, 11. Ancryodelloides trigonicus Bischoff and
    Sannemann (P). 3, 4, top and basal views of UCR
    8530/14; 5, 6, and 11, top, side and basal views of UCR
    8530/13 from MC 3C.

Figs. 9, 10. Ancryodelloides transitans (Bischoff and
    Sannemann) (P). 9, top view of UCR 8539/10 from MC 7;
    10, top view of UCR 8530/9 from MC 3C.

Figs. 12-17. Pandorinellina? cf P.? boucoti (Klapper) (P).
    12, 13, top and side views of UCR 8576/6 from MC 20C;
    14-17, top, basal, side and other side views of UCR
    7346/1 from Tor Limestone, Toquima Range, Nevada.

Figs. 18, 19, 21-24. Ancryodelloides spp. (O) element. 18,
    19, top and side views of UCR 8532/2 from MC 4A; 21,
    side view of UCR 8532/3 from MC 4A; 22-24, side, top
    and other side views of UCR 8534/1 from MC 5A.

Fig. 20. Ancryodelloides kutscheri Bischoff and Sannemann,
    (P). Top view of UCR 8532/5 from MC 4A.

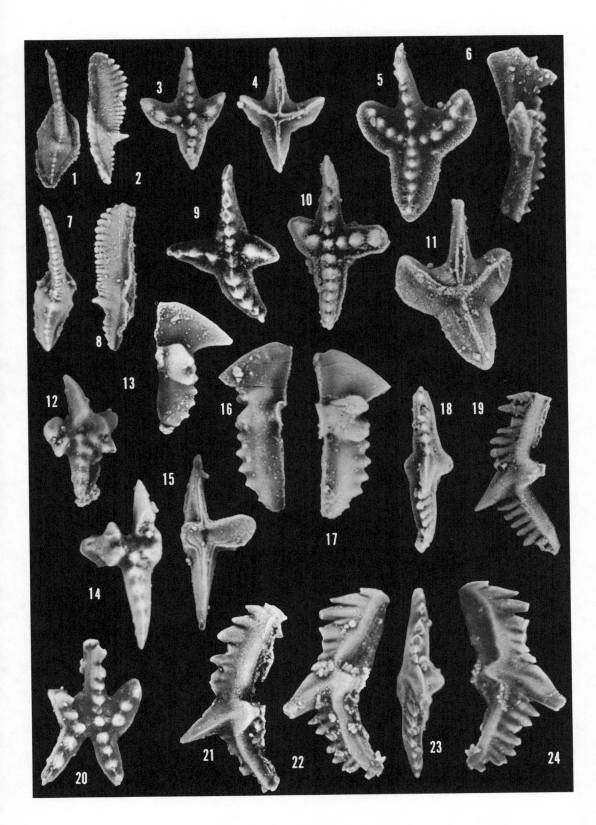

Plate 4--All figures approximately 30X.

Figs. 1-3. Ancryodelloides sp. (O), probable A. delta-
eleanorae (O) element, side, basal and other side
views of UCR 8534/2 from MC 5A.

Figs. 4-6. Ancryodelloides eleanorae (Lane and Ormiston)
(P), top, basal and side views of UCR 8530/15 from MC
3C.

Figs. 7-9. Ancryodelloides delta (Klapper and Murphy) (P),
basal, side and top views of UCR 8530/6 from MC 3C.

Figs. 10-12. Ancryodelloides specimen with characteristics
of A. eleanorae, A. omus and A. transitans (P), top,
basal and side views of UCR 8529/2 from MC 3A.

Figs. 13, 19. Pandorinellina optima (Moskalenko) (P), side
views of UCR 6205/2 from COP II 65.5 and UCR 6268/1
from COP IV 414, respectively.

Figs. 14-18. Ancryodelloides delta (Klapper and Murphy)
(P). 14, 15, side and top views of UCR 8903/8 and 16-18,
top, basal and side views of UCR 8903/7, respectively,
from MC 3.

Figs. 20-30. Ancryodelloides limbacarinatus n. sp. (P).
20, 25, 26, basal, side and top views of the holotype,
UCR 8535/1 from MC 5B; 21, 22, top and side views of
UCR 8539 from MC 7; 23, 24, top and side views of UCR
8780/ 3 from SP VII 365; 27, top view of UCR 8780/2 from
SP VII 365; 28-30, top, basal and side views of UCR
8780/1 from SP VII 365.

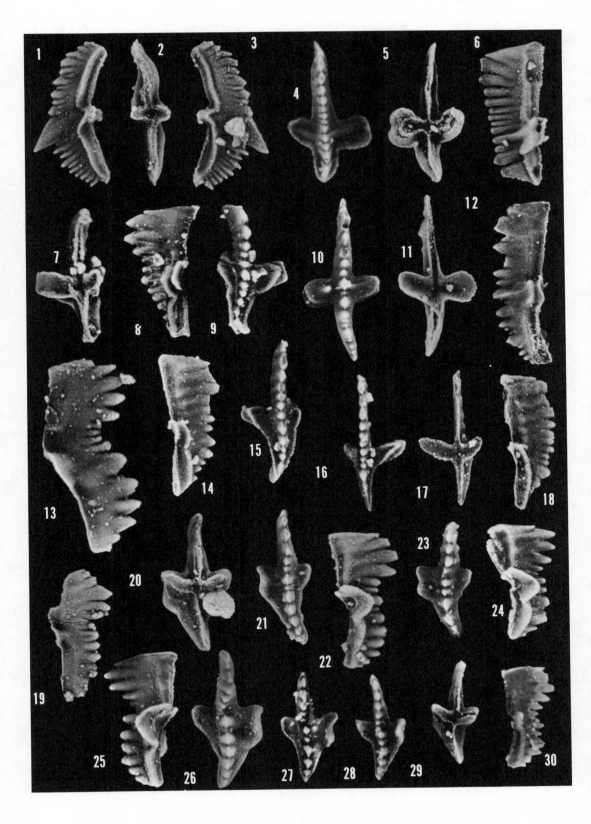

Plate 5--All figures approximately 30X.

Figs. 1-15. <u>Amydrotaxis</u> <u>sexidentata</u> n. sp. (P). 1, 2, side

and top views of the holotype, UCR 8524/1 from MC 2G +
7cm; 3, side view of UCR 8982/1 from SP I 1672 feet;
4-6, basal, top and side views of UCR 7405/3 from BC II
713.5 feet; 7, side view of UCR 7408/2 from BC II 1685
feet; 8, 9, side and top views of UCR 6297/1 from COP V
121 feet; 10-12, top, basal and side views of UCR 8524/3
from MC 2G + 7 cm; 13-15, top, side and basal views of
UCR 6237/2 from COP IV 0 feet.

Figs. 16-19. <u>Amydrotaxis</u> <u>sexidentata</u> n. sp. (O), side and
basal views of UCR 7007/4 and 7007/6 from SP VII 181
feet.

Figs. 20-22. <u>Amydrotaxis</u> <u>sexidentata</u> n. sp. (B_1), side
views of UCR 7007/7, 8977/2, and 8978/1 from SP VII
181, SP I 1617, and SP I 1658 feet, respectively.

Figs. 23, 24. <u>Amydrotaxis</u> <u>sexidentata</u> n. sp. (B_3),
posterior views of UCR 8982/2 and 8977/3 from SP I 1672
and SP I 1617 feet, respectively.

Figs.25-27. <u>Amydrotaxis</u> <u>sexidentata</u> n. sp. (B_2), side
views of UCR 7007/8, 8977/1 and 8978/2 from SP VII 181,
SP I 1617, and SP I 1658 feet, respectively.

Figs. 28, 33. <u>Pedavis</u> <u>biexoramus</u> n. sp. (I), top views of
UCR 9035/2 and 9035/1, holotype, respectively, from Ikes
Canyon, Toquima Range, Nevada.

Figs. 29, 34. <u>Icriodus</u> n. sp. G. of Klapper (I). 29, top
view of UCR 9021/1 from Ikes Canyon, Toquima Range,
Nevada; 34, top view of UCR 8523/2 from MC 2G.

Figs. 30, 35. <u>Icriodus</u> sp. (I), top views of UCR 9020/1
and 8977/10 from Ikes Canyon, Toquima Range, Nevada and
SP I 1617 feet, respectively.

Fig. 31. <u>Icriodus</u> <u>steinachensis</u> Al Rawi, beta morph, top
view of UCR 6211/34 from COP II 163 feet.

Fig. 32. <u>Icriodus</u> <u>claudiae</u> Klapper and Johnson, beta morph,
top view of UCR 6212/8 from COP II 177 feet.

Fig. 36. <u>Icriodus</u> <u>steinachensis</u> Al Rawi, eta morph, top
view of UCR 6232/1 from COP IIA 47 feet.

Fig. 37. <u>Icriodus</u> <u>claudiae</u> Klapper and Johnson, alpha
morph, top view of UCR 6211/26 from COP II 163 feet.

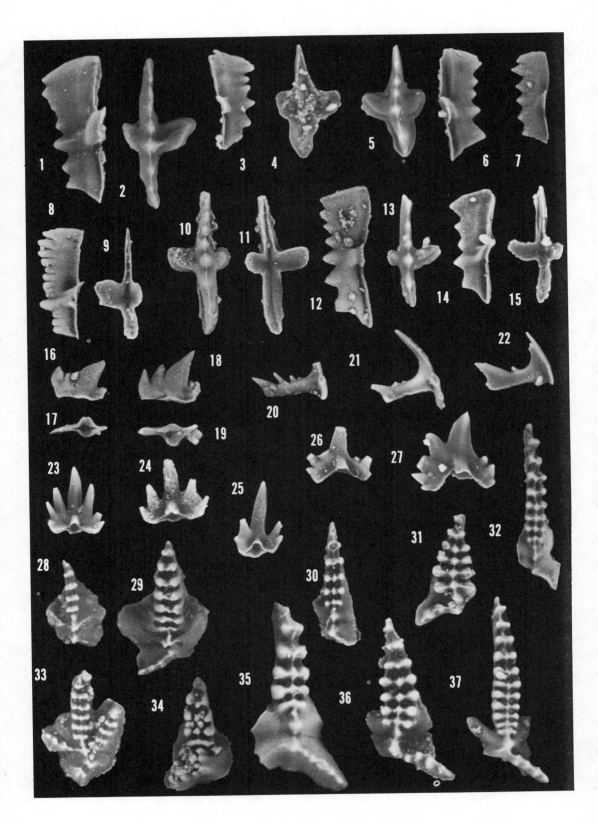

Plate 6--All figures approximately 30X.

Figs. 1-13. Erika divarica n. sp. 1, 2, (Pa) side and other
    side views of UCR 8993/6 from SP VII 348 feet; 3, (Pa),
    side view of UCR 8776/4 from SP VII 359 feet; 4, (Pb),
    side view of UCR 8776/2 from SP VII 359 feet; 5, 9, (Sb)
    side view of 8765/3 from SP VII 250 feet and upper
    oblique view of UCR 8766/3 from SP VII 260 feet,
    respectively; 6, 7,(Sa), side views of UCR 8767/5 and
    UCR 8765/12 from SP VII 274 and SP VII 250 feet,
    respectively; 8, (Sc) side view of UCR 8771/4 from SP
    VII 337 feet; 10, 11, (Sb or Pb), side views of UCR
    7410/1 and 8256/1 from COP IV and Wenban Peak, Cortez
    Range, Nevada, respectively; 12, 13, (Sc), side and
    oblique anterior views of UCR 8230/1 from Wenban Peak,
    Cortez Range, Nevada.

Figs. 14-16. Pedavis brevicauda n. sp. (I). 14-16, top and
    basal views of UCR 6211/1, holotype, and top view of UCR
    6211/2 from COP II 163 feet; 16, top view of UCR 7343/1
    from Willow Creek, northern Roberts Mountains, Nevada,
    in the basal bed of the McColley Canyon Formation.

Fig. 17. Pedavis brevicauda n. sp. (I), top view of UCR
    6249/22 from COP II 163.

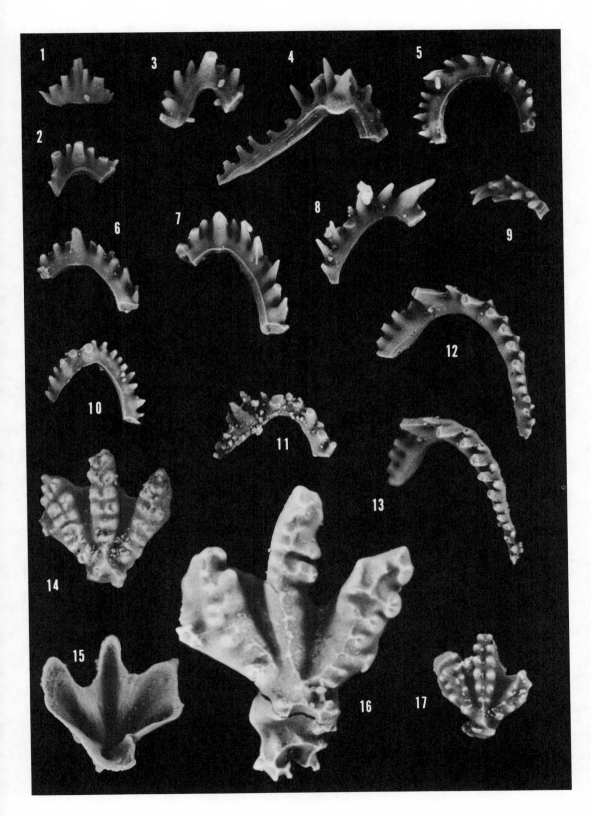

Plate 7--All figures approximately 30X.

Figs. 1, 9-11. *Pedavis* cf. *P. breviramus* n. sp. (I). 1,
   specimen with broken process, top view of UCR 8997/10;
   9-11, (S_1), top and two side views of UCR 8997/11
   from U Topolů section, Radotín Valley, Czechoslovakia.

Figs. 2, 3, 6, 12. *Pedavis breviramus* n. sp. (I), top views
   of UCR 8253/1 from Wenban Peak, Cortez Range, Nevada,
   UCR 6251/2 from COP IV 164 feet, UCR 8767/9, holotype,
   from SP VII 274 feet, and UCR 8785/1 from SP VII 391
   feet, respectively.

Fig. 4. *Pedavis* cf. *P. breviramus* n. sp. (I), top view of
   UCR 8765/14 from SP VII 250 feet.

Fig. 5. *Pedavis* cf. *P. breviramus* n. sp. (I), top view of
   UCR 8993/1 from SP VII 348 feet.

Figs. 7, 8. *Pedavis* sp. (M_{2a}), side views of UCR 8765/16
   from SP VII 250 feet.

Figs. 13, 20. *Pedavis pesavis* sp. (M_{2a}), side views of
   UCR 6272/4 from COP IVA 8 feet.

Figs. 14, 15. *Pedavis* sp. (S_1), side views of UCR 7345/2
   from COP IV 371 feet.

Figs. 16-19. *Pedavis* sp. (M_{2d}), two side, posterior and
   anterior views of UCR 6268/6 from COP IV 414 feet.

Fig. 21. *Pedavis* sp. (M_{2b}), side view of UCR 8767/7 from
   SP VII 250 feet.

Figs. 22, 23, 24. *Pedavis* sp. (M_{2c}), side, posterior, and
   other side views of UCR 6272/3 from COP IVA 8 feet.

Figs. 25, 26. *Pedavis* sp. (M_{2d}), side and anterior views
   of UCR 7404/22 from COP IV 404 feet.

Fig. 27. *Pedavis* sp. (M_{2c}), side view of UCR 8993/8 from
   SP VII 348 feet.

Fig. 28. *Pedavis* sp. (M_{2b}), side view of UCR 8993/7 from
   SP VII 348 feet.

Figs. 29-31. *Pedavis* sp. (M_{2a}), anterior, side, posterior
   and other side views of UCR 7407/23 from COP IV 404
   feet.

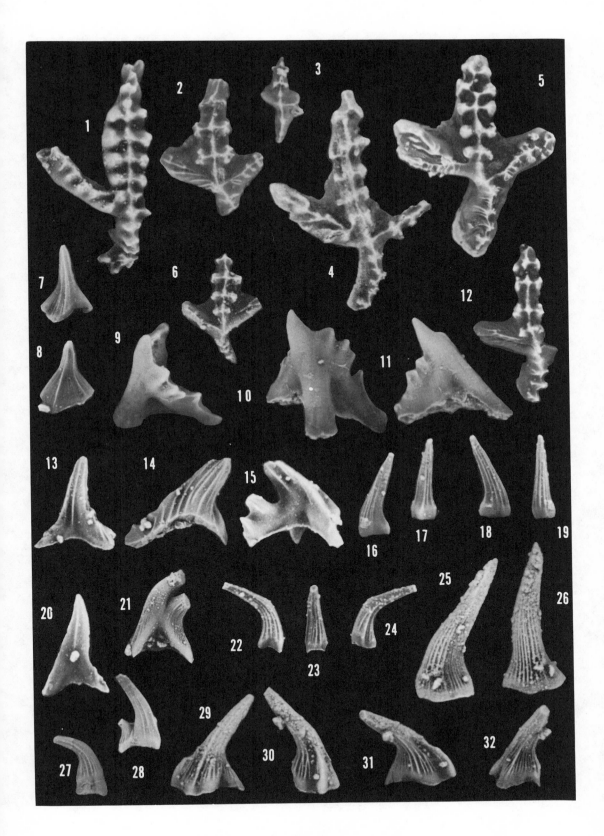

Plate 8--All figures approximately 30X.

Figs. 1, 2, 4-6, 9, 11, 12. _Pedavis mariannae_ Lane and
    Ormiston (I). 1, 2, 4, 6, top views UCR 7344/2, 4, 3,
    and 1, respectively, from COP II 378 feet; 11, basal
    view of specimen in fig. 1; 5, 9, top and basal views of
    UCR 8973/13 from COP II 295 feet; 12, top view of UCR
    6220/1 from COP II 335 feet.

Fig. 3. _Pedavis_ sp. (I), top view of UCR 8973/2 from COP II
    295 feet.

Fig. 8. _Pedavis latialata_ (Walliser) (I), from the Boot-
    strap Mine Area, north of Carlin, Nevada, for comparison
    with _Pedavis breviramus_ n. sp.

Figs. 7, 10. Pedavis sp. (I), top and basal views of UCR
    7409/1 from COP IV 413 feet.

Fig. 13. _Pedavis breviramus_ n. sp. (M_{2b}) side view of UCR
    8993/7 from SP VII 13B.

Figs. 14, 15. _Pedavis breviramus_ n. sp. (M_{2c}), side views
    of UCR 8993/14 and 8993/8 from SP VII 13B.

Figs. 16, 17. _Pedavis breviramus_ n. sp. (M_{2d}), side and
    other side views of UCR 8993/3 from SP VII 13B.

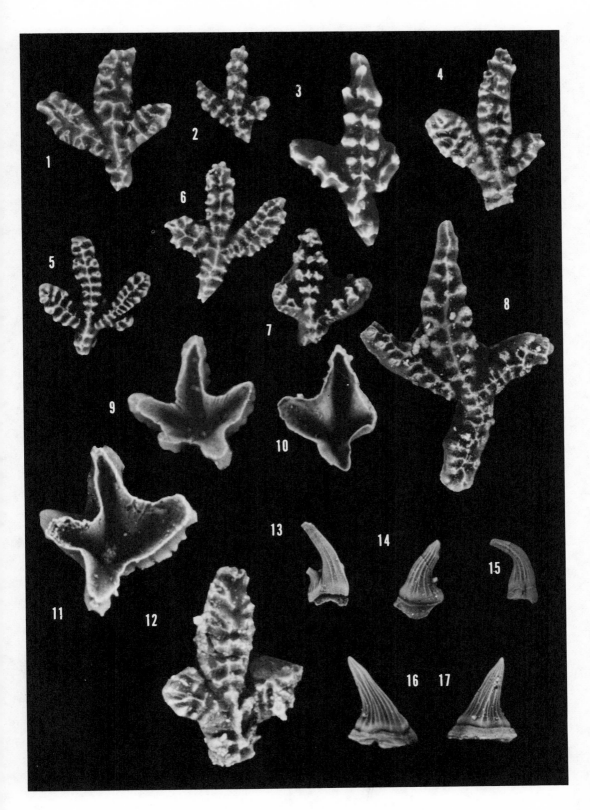